Math Survival Guide
Tips for Science Students

Jeffrey R. Appling

Director of General Chemistry
Clemson University
Clemson, South Carolina

John Wiley & Sons, Inc.
New York • Chichester • Brisbane • Toronto • Singapore

ISBN 0-471-03103-8

Printed in the United States of America

10 9 8 7 6 5 4 3 2 1

to Paula

What type of calculator should you buy before you start your first year of college science? Should you buy a computer? How do you estimate a logarithm without using a calculator? What study technique has raised exam scores by more than 30 points? In this book I answer these questions, and provide other ideas and strategies that might help you in your first college science courses.

Doing math, even simple math it seems, is like speaking a foreign language. If it has been a while since you spoke the language, you can expect to be a little rusty. Students sometimes need to brush up math skills in preparation for a first year science course. This book is meant as a review to help those students, and students presently in science courses who need a quick reference to basic math techniques. It is not a text, and it does not cover every aspect of useful mathematics. Rather, it is a "survival guide," where basic ideas have been gathered together for discussion. Sometimes that discussion is brief and efficient. Sometimes the discussion has more depth in order to get across important concepts. Either way this book should help you if you feel a little shaky about your math skills. In addition, I have included study tips and insights that could help you in all of your science courses. By exploring these ideas you will develop skills that you need to become a successful science student.

If you are preparing to take your first college level science course you should read this book from the beginning. After the term progresses you can use this book as a resource to help remind you of forgotten methods, and learn tips to help you "invent" your own shortcuts to solve problems. You probably know much of the math contained in this book. What I have tried to write is a summary packed with hints that will bring your math skills up to speed efficiently.

This book starts with the basics of numbers and their manipulations, including ratios, powers, roots, logarithms, scientific notation, and significant figures. With numbers under control, the reader can advance into the territory of equations, units, and graphing. A short review of trigonometry and geometry precedes chapters aimed at improving problem solving and study skills. The reader is introduced to helpful methods (estimation techniques, for example) that also enhance understanding. These later chapters target ways to study more efficiently and with better retention. I have included these chapters to make this survival guide more than a math summary. The suggestions in this book could advance your development of good study habits. I compiled these tips by observing students—the good students. Finally, I have included exercises in each chapter with answers in the appendix. Do these problems to strengthen your skills and try out new things that you learn in the chapters.

I hope this book comes in handy during your first year science course. At the end of that year you will benefit from books written at a higher level, particularly those that cover calculus and matrices. An excellent choice for science and engineering majors is "Mathematical Methods in the Physical Sciences," by Mary Boas (John Wiley & Sons, 1983). This book is written at the sophomore level, yet covers topics that will carry you throughout your career as a scientist.

It has been a pleasure working with my editor, Eric Stano. His enthusiastic encouragement, patient guidance, and timely weather reports were critical to the success of this project. I would like to thank Nedah Rose at John Wiley & Sons for her encouragement, and for helping me bring an embryonic idea to maturity. Special appreciation goes to my wife Paula (my "other editor") for her help and support during this project. Finally, I am indebted to the following reviewers for their insights and helpful criticisms:

John Barach, Vanderbilt University
Tom Colbert, Augusta College
Robert Lloyd Ernst, Southwest Missouri State University
Ralph T. Grannan, College of the Desert
George A. Page, Central Connecticut State University
Barbara Rainard, Community College of Allegheny County
Ann Ratcliffe, Oklahoma State University
Dennis F. Ragan, Green River Community College
Donald Ruch, Transylvania University
Theodore Sakano, Rockland Community College
Samuel T. Scott, Westfield State University
Thomas M. Snyder, Lincoln Land Community College
Gabriel Sunshine, New York Institute of Technology
Trudie Jo Slapar Wagner, Vincennes University

I wrote this book for students, and I am genuinely interested in what students think about it. Let me know how you feel about "Math Survival Guide." If you have any pet tips or tricks that I can add, send them to me via snail-mail (U. S. mail) or electronic mail at Appling@Indigo.Clemson.Edu. I look forward to hearing your thoughts.

Jeffrey R. Appling
Clemson, South Carolina
January, 1994

CONTENTS

Why shouldn't you buy the most powerful calculator you can afford? Exactly what should your calculator do?

What are integers, fractions, irrational numbers, and complex numbers? Do you need to know the values for **pi** and **e**?

What is the difference between a ratio and a fraction? What do you need to know about percentages that is so obvious you might not see it when you need it? How do you calculate ppt, ppm, and ppb? What is a proportion good for?

Why use powers? What do you do with a negative exponent? What does a fractional exponent mean? How can you learn to multiply and divide exponents quickly?

What is a logarithm? What is the difference between a common log and a natural log? How are logs and exponents related? If your calculator has no **antilog** key, how do you calculate one?

When and how do you use scientific notation without a calculator? What is it about scientific notation that your calculator can't handle? What is the most common type of mistake made when using scientific notation?

What are the rules for figuring significant digits? How is reporting significant figures in a calculation involving logarithms different?

What is the most important thing to remember when manipulating an equation? How can you manipulate equations quickly? How do you solve quadratic equations? When should you reverse the direction of an inequality?

Why should you care about units? What are SI units? What unit conversions should you commit to memory?

How do you graph data in the most effective and presentable way? What dangers lurk in extrapolation of data? How can you justify throwing out a data point? How do you plot the best line through your data points?

How are triangles and circles related? What do Oscar and Pythagoras have in common? Your calculator distinguishes between degrees and radians. Do you? Can you calculate perimeters, areas, and volumes of regular shapes?

What is a good, basic approach to solving problems? Why is cramming for an exam ineffective? Homework is such a bother, is it necessary?

How can your test scores benefit by learning how to estimate? How can you estimate solutions to problems? How do you estimate roots and logarithms?

What is the best color pen to use to highlight your textbook? If your instructor is a snore, why should you go to class? What is the best way to read your textbook? How much time should you spend in out-of-class studying? When should you get a tutor? What is the *biggest* freshman mistake? What two studying techniques have dramatically improved students grades (30 points!)?

Do you *need* a computer? Your heart is set on buying a computer—what should it do? What is the Internet, and why is it important for students?

Math Survival Guide
Tips for Science Students

Your Calculator: Friend or Foe?

It sure seemed like a good idea at the time. To tackle the imposing mountain of science you bought a state-of-the-art calculating machine. A miniature computer. To be your helper, your companion, your weapon against the sinister forces of math. It had buttons. *Lots* of buttons. You named it Hal.

But here you are sitting in the middle of an exam, and your buddy has deserted you. It now speaks a foreign language. You punch in [2 + 2]. It answers quietly: '11'. You try division. It blinks at you. Nothing you try works, not even turning it off and turning it back on (which took some time since it doesn't have an on-off switch—you had to wait until **it** decided it wanted to shut off). You've wasted about ten precious minutes, and now you have to grind through the rest of the test by hand. You can't do the logarithm problems.

After the test you approach your instructor. "My calculator died," you report. As brief discussion ensues, you're hoping for some special consideration. Your instructor probes you about your problems. You reveal grudgingly that your space-age sidearm actually works, after a fashion, but that you can no longer get its attention. Your instructor picks up the calculator, and hands it back after a few keystrokes. You check again. '4' is the response. Your instructor walks away with the exams. "Mode key" is all she says. Your heart sinks.

The above scenario need not occur (and unfortunately it does all too often). Locking up your calculator during an exam happens for only one reason—you don't really know how it works. It is easy to hit the wrong key during a calculation. Sometimes this can have disastrous results, especially if you don't know how your calculator works well enough to know how to undo the damage.

It is now easy to afford a powerful handheld calculating machine. Some have the capabilities found only in desktop models of a few years ago. A reasonable sum brings you mathematical wizardry in a box. But the average student does not have the time to plow through the 100 page owner's manual, so most people wing it with their new toys. The math that you will use in most first year science courses is relatively simple. With the exception of roots, logs, and antilogs you will be able to do the calculations by hand. But the sophisticated calculator goes beyond this—way beyond what you really need. Using one of these button-burdened beasts is like peeling a grape with a machete. You are out-gunning the problem.

The solution is easy. No matter how tempting they are, with their graphics screens and alphanumeric memories, do not buy a sophisticated calculator UNLESS you are willing to really learn how it works. Instead, buy a calculator more suited to your real, current needs. Quite a few models will do the job for less than twenty dollars. Check out the nearest "office warehouse" for a nice selection.

So what functions does your "nerd-o-lator" really need? They all have the big four: addition, subtraction, multiplication, and division. You will need exponential notation (so the freebie from the bank won't really cut it) and at least one memory location. Memories are sometimes shown as **STORE/RECALL** keys. A key to allow changing the sign (from plus to minus and vice versa, ± or **CHS** for instance) comes on most models. An **IN-VERSE** key (often **INV**) will allow you to do the reverse of certain functions. The **IN-VERSE** function is essential when doing trigonometric calculations like sines, cosines, etc. A key labeled **1/x** will allow you to invert numbers and fractions. For example, the easy way to get the decimal value of "one fifth" is to enter the number **5** followed by the **1/x** key (two keystrokes) to yield **0.2**, which is the same result you get from the four keystrokes: [1 ÷ 5 =]. Throughout this book we will use bold brackets ([]) to show a sequence of calculator keystrokes.

For simple trigonometry, the function keys **SIN, COS,** and **TAN** (coupled with their inverses) are sufficient. A calculator equipped with these functions will also have the capability to do calculations either in degrees or radians, and will have a key to switch between the two. It will also be able to quickly bring up the value of **pi** (π = 3.14159265358979323846 ...).

You will need functions based on exponents ("powers") and their inverses. For instance, to easily **SQUARE** a number the x^2 key is handy. The **SQUARE ROOT** key, $\sqrt{}$, will do just the opposite. For powers other than two the key labeled x^y allows any combination. You enter x first, followed by the power to which x is raised. For example, suppose you need the fourth power of ten—10^4. You would enter [10 x^y 4 =], which would yield the desired answer—**10,000**. Taking roots other than square roots is just as straightforward. The cube root (1/3 root) of one million would be: [1000000 $x^{1/y}$ 3 =], which yields the result **100**. The $x^{1/y}$ key allows for fractional exponents.

The powers of ten are common in scientific calculations, so a 10^x key is useful. To determine exponents for numbers in base 10 you use a function called the logarithm. A **LOG** key designates this function. Other useful operations in science are done in "base **e**." Power functions with **e** are done with an e^x key, and the opposite function, the natural logarithm, uses the **LN** key. We will discuss powers and logarithms in more detail in Chapters 4 and 5.

The previous examples are based on the types of keystrokes one might use with most calculators on the market. These calculators will have an **EQUALS** key, =. In this book all examples will be given using this type of logic. Some other calculators, most notably those manufactured by Hewlett-Packard, use a different type of logic, which changes the way you input keystrokes to perform a calculation. The difference is slight, so if you have one of these machines you will easily follow examples in this book. Read your owner's manual to get a handle on the difference.

Some students buy a complicated calculator with the idea that it will be all they need for their college years, from the easiest to the hardest classes. **This is a mistake**. Buy the simple, cheap calculator first. Your early years will be smooth and you can let the technology catch up to your advanced courses. When you need a sophisticated machine it will be cheaper than if you bought it years before, when you really weren't ready to exploit its capabilities. And you'll find that your original $15 model is still useful. It might be all that you ever need (it was for me, I used the same one for nine years of college and graduate school!).

In this book we will explore some of the manipulations available on your calculator, including some shortcuts. The aim of this book is to review the math necessary to do well in first year science courses. Math is the "language" used for solving problems. The next chapter will introduce some "letters" of this language—numbers.

Chapter 1 Summary

- Buy an inexpensive, simple scientific calculator.

- Look for these keys: + - × ÷ =

    EXP (or **EE**) for exponential notation

    STO (or **M**) for memory

    $\pm$ **INV** 1/x

    π **SIN COS TAN**

    x^2 $\sqrt{}$ x^y $x^{1/y}$ **10^x LOG** e^x **LN**

- If you do buy a powerful calculator, *learn how to use it!* A mistake can mean a melt-down (watch those pesky modes).

Practice Exercises Chapter 1

Try these problems to exercise your calculator skills. Calculate an answer for the quantity shown.

1. $\dfrac{3.53+6.20}{3.53-6.20}$

2. $\dfrac{1}{4(3.5)-6.0}$

3. $(4.12)^3 - 2.33 \ \log(6.2-2.6)$

4. $5+\sqrt{(5)^2 - 4(1)(6)}$

5. $\dfrac{4.7\times10^{-3} + 2.3\times10^{-3}}{6.5\times10^{-7}}$

6. $x^2 - 5x + 6$ for $x=3$

7. $\dfrac{(2x)^2}{(0.10-x)(0.10+x)}$ for $x=5.0\times10^{-2}$

8. $e^{-\left(\frac{E}{RT}\right)}$ for $E = 5000$, $R = 8.31$, and $T = 298$

9. $\sin(\Pi/2)+\cos(\Pi/2)$

10. $\tan^{-1}(1.00)$ in degrees

Numbers

Now that you have a good, dependable calculator you'll need some numbers to put into it. Even though success in science courses will depend on your grasp of the concepts, you will often have to show your abilities through the use of numbers. So it will pay you to master them. In this chapter you will get re-acquainted with some old friends.

By the age of three you could probably count to ten. Those counting numbers were all positive (unless you were a very strange child), but some time later you learned the concept of **zero**. It was probably much later that you added the negative numbers to complete the number line:

$$-\infty \text{ ... -4, -3, -2, -1, 0, +1, +2, +3, +4 ... } +\infty$$

These numbers are **integers** or **whole numbers**. The symbol ∞ represents **infinity** (an interesting concept that is difficult to fathom whether you are 3 or 300).

At some point you learned about money. There were quarters, nickels, dimes, pennies—all parts of a dollar. With a money system based on one hundred pennies in a dollar, you soon learned about the **decimal** system. The price tag of $2.98 meant 298 pennies. Using the decimal point, you could easily show fractions of dollars. Now there were other "numbers" between the integers.

A **fraction** represents a value between two **whole numbers**. Fractions are written as one integer divided by another. To convert a fraction to a decimal number, simply divide the upper number (**numerator**) by the lower number (**denominator**). A quarter is, of course, $\frac{1}{4}$. Dividing **1** by **4** yields **0.25**. Fractions greater than one can be represented as either a straight ratio (more on ratios in Chapter 3) or a **compound fraction**. For example, $\frac{23}{8}$ is equal to the compound fraction $2\frac{7}{8}$. Learn to convert compound fractions quickly. To convert $2\frac{7}{8}$ the numerator will be $7 + (8 \times 2) = 23$. The new denominator is identical to the old one in the compound fraction. You can also think of $2\frac{7}{8}$ as $2 + \frac{7}{8}$, or $\frac{16}{8} + \frac{7}{8} = \frac{23}{8}$, since 2 is equal to $\frac{16}{8}$.

Numbers in science are often decimal, but are not restricted to two places after the decimal. For instance, if you measured a mass to the nearest milligram it might have the value 2.137 grams. The significance of how many positions are occupied in a number (what's the difference between 2.137 and 2.13700?) will be covered in detail in Chapter 7. In Chapter 6 we will discuss **scientific notation**, the most common and useful way of presenting very big or very small numbers.

Most numbers that you will encounter have specific values. However, there are two important exceptions. The first you are probably familiar with already—**pi**. **Pi** has a value close to 3.14159265358979, but it is an **irrational number** so the decimal equivalent has an infinite number of places. You will encounter **pi** in calculations using angles in geometry and trigonometry (covered in Chapter 11). The other common irrational number is **e**. Base **e** is important in scientific calculations; it is the base of the **natural logarithm** (more on logarithms in Chapter 5). In solving problems you will sometimes need to use an approximation for **e**, which is 2.71828.

The numbers discussed so far, with the detail provided in subsequent chapters, will cover common calculations that you will encounter in your first few years of science course work. Eventually you will need to manipulate **complex numbers**, and they are mentioned here for completeness. A complex number is a combination of two parts. The first part is a **real number**, like the integers, decimals, and fractions discussed above. The second part consists of a **real** coefficient times the square root of -1: *i*. The complex number is thus written as two parts like **2 + 3*i***, or more generally **a + b*i***.

Chapter 2 Summary

- Numbers commonly used in scientific calculations are **integers**, **decimals**, and **fractions**.

- The **irrational numbers pi** and **e** are used in special types of calculations.

- Most calculations are conducted with **real numbers**, although **complex numbers** are important in advanced scientific work.

Practice Exercises Chapter 2

1. Convert these compound fractions to simple fractions and decimal numbers.

a. $5\,^5/_9$ c. $3\,^5/_{11}$

b. $2\,^{23}/_{32}$ d. $-4\,^2/_7$

2. Convert these simple fractions to compound fractions and decimal numbers.

   a. $\dfrac{13}{8}$ c. $-\dfrac{56}{26}$

   b. $\dfrac{305}{137}$ d. $\dfrac{256}{16}$

3. What are the decimal equivalents for these common simple fractions?

   a. $\dfrac{1}{8}, \dfrac{3}{8}, \dfrac{5}{8}, \dfrac{7}{8}$

   b. $\dfrac{1}{3}, \dfrac{2}{3}, \dfrac{4}{3}, \dfrac{5}{3}$

4. Convert these decimal numbers to simple fractions and compound fractions (see #3 for help).

   a. 13.375   c. $-4.3\overline{3}$

   b. 7.625   d. $3.6\overline{6}$

5. Use your calculator to determine values for the irrational numbers π and e.

Ratios

Think about the first day of one of your classes—History, for example. It's natural to look around and learn about your classmates. Let's say you count 60 women and 40 men (including yourself). When asked about your class by your nosey roommate, you might respond "There are 60 women and 40 men." But there are other ways to express this as well. We often use a **ratio** to relate the amounts of things.

To retain the original numbers of students in your class, you might tell your roommate "The male-to-female ratio is 40 to 60." You could write this **40:60**, using a colon ("**:**") as the symbol to indicate a ratio. It would be more common, however, for you to mentally reduce the ratio of 40:60 by dividing by a common factor. You could relate the same information with the ratio **4:6** (dividing by 10), or more simply as **2:3** (dividing by 2 again, or dividing the original ratio by 20). Your roommate then would know that there were three women students for every two men.

Ratios are useful in science. They can give relations between items in a quantitative way. A common use for ratios is as "conversion factors" (unit conversions will be discussed in detail in Chapter 9). For example, you know that there are 12 inches in one foot. The ratio of inches:feet is therefore 12:1. You can also represent the same thing as:

$$\frac{12 \text{ inches}}{1 \text{ foot}}$$

where the ratio symbol has been replaced by the familiar division line. You might read this as "there are 12 inches per foot." By including the units of inches and feet the method is more compact. You don't need to preface the ratio of 12:1 with the explanation "inches:feet." We will use this format in the chapter on units and their conversion.

A **proportion** is an equation formed by equating two ratios. Using the example from above, 2:3 = 40:60 represents one proportion that can be stated about your history class.

While this is a proper proportion, you will see it most often written in the format:

$$\frac{2}{3} = \frac{40}{60}$$

Proportions are useful when you know a ratio, but would like to scale up or down. You might ask the question "If I make $6.00 per hour, how much money will I make in a 40 hour week?" This can be expressed as a proportion with an unknown quantity that you want to know:

$$\frac{\$6}{1 \text{ hr}} = \frac{?}{40 \text{ hr}}$$

This equation can be satisfied when the unknown quantity is equal to $240, since $6:1 = $240:40. Of course, in your head you arrived at the answer by multiplying 40 hours times your rate of $6 per hour. When you compute how much you will make in a given time, you are solving a proportion. We will work more with equations in Chapter 8, and use proportions to help convert units in Chapter 9.

One very important issue that involves equations and proportions is the idea of **proportionality**. Consider the density of objects, whether they are "heavy" or "light." The density D is determined by the object mass (m) and its volume (V):

$$\text{Density} = \frac{\text{mass}}{\text{Volume}}$$

$$D = \frac{m}{V}$$

For a given volume (size), as the object mass increases so does its density—it gets "heavier." This type of relation (D increases when m increases) is a *direct* proportionality. The density and mass are said to be **directly proportional**. The relationship between density and volume is just the opposite. For a given mass, as the volume increases the density *decreases*. You would say that a coin with mass 2.2 kilogram (1 pound) is heavy, whereas a 2.2 kilogram piano would seem extremely light. Density and volume are **inversely proportional**, since D decreases as V increases. This behavior happens because D is in the numerator of the left side of the equation and V is in the denominator of the right side. As V gets larger (coin → piano), $\frac{m}{V}$ gets smaller (you are dividing by a larger number) and hence D decreases.

Let's go back to your first day of history class. What if your nosey roommate doesn't phrase the question as "What is the male-to-female ratio?" An equivalent question could be "What fraction of your history class is women?" This is a request for the same information, but in a different form. Recall that the ratio was 60/40, women/men. The **fraction** of women is equal to the number of women related to *the sum* of all the students in the class (presumably only in the categories "men" and "women"—no "other"). You must relate the number of women to the addition of women plus men:

$$\frac{60 \text{ women}}{60 \text{ women} + 40 \text{ men}} = \frac{60 \text{ women}}{100 \text{ students}} = 0.60$$

However, it is unlikely that your roommate would understand your response as "Zero point six."

How do you communicate that 6 out of 10 students are women? Simply use a **percentage.** You will naturally convert 60 women per 100 students (total) to **60%.** Therefore, the number 60% is equivalent to 60/100, which is 0.60. To find the percentage of something, first find the fraction and multiply by 100%. So the fraction of women in your history class is 0.60 and the percentage of women is 60%. The fraction of men is 1 - 0.60 = 0.40, and the percentage of men is 100% - 60% = 40%.

Any number represented as a percentage shows the "parts per one hundred." Most totals won't be 100, of course, so you must first find the fraction and then convert to percentage. For example, if your biology class has an enrollment of 195 and 4 students graduated from your high school, you and your high school friends make up about 2% of the class: 4/195 = 0.0205, 0.0205 × 100% = 2.05%.

Remember that any percentage can be converted to its decimal equivalent for use in calculations. For example, if 13% of your Calculus class made an "A" on the first exam, and 54 students are in the class, you can determine the number of students that made an "A": 13% of 54 = (13%/100%) × 54 = 0.13 × 54 = 7.02. So seven students made an "A". Note that to use the percentage on your calculator you had to convert it to the decimal equivalent.

An important consequence of using percentages is that the sum of all percentages must equal 100%. Think of 100% as 100/100, which is equal to 1. All parts of a whole must sum to equal 1. You can use this to get you out of a jam sometimes. It may appear that a problem is missing information, when in fact you can determine the missing number by remembering that the sum of all parts must equal 1 (or all percentages must sum to 100%). Consider the following problem:

> Analysis of a hydrocarbon containing carbon, hydrogen, and oxygen yields 59.99% carbon and 4.48% hydrogen. What is the empirical formula for this compound?

In order to begin this problem you need the percentage of oxygen, which is not given. Don't panic, you can get it easily:

59.99% C + 4.48% H + **?% O** = 100% (all percentages must sum to 100%)

%O = 100% - (59.99% C + 4.48% H) = 100% - 64.47% = **35.53% O**

(This compound is aspirin, for those of you playing along at home).

Say your fruit fly colony has 14 mutants in a total of 1272 flies. You determine the fraction of mutants as 14/1272 = 0.011 and convert to percentage by multiplying the answer by the number 1 (remember, you can always multiply anything by the number 1 without changing its value). In this case you might choose to represent the number 1 as 100%: 0.011 × 100% = 1.1% mutants. What about another expression of the number 1? How about in "parts per thousand," also known as **ppt.** Just as 100% is equal to 1, 1000 ppt (1000/1000) is equal to 1. This allows you to represent your mutants in another way: 0.011 × 1000 ppt = 11 ppt mutants. In other words, for every group of 1000 flies, 11 will be mutants. Both forms are acceptable—and equal to one another. You can use "parts per million," **ppm**, in a similar fashion. In this format, 0.011 × 1,000,000 ppm = 11,000 ppm. This number is too bulky, so you might choose to use percentage or ppt for a more streamlined result. Parts per million and parts per billion (**ppb**) are typically used for small concentrations of species in water (toxins for example).

Chapter 3 Summary

- **Ratios** relate amounts of two things **a** and **b** in the format **a:b** or **a/b**.

- A **proportion** is an equation where ratios are equated.

- A **fraction** relates one thing to the whole. Common ways of expressing fractions include **percentage (%)**, **parts-per-thousand (ppt)**, **parts-per-million (ppm)**, and **parts-per-billion (ppb)**.

- The sum of all fractions must equal **1**. For example, the sum of all percentages must equal 100%.

Practice Exercises Chapter 3

1. Simplify these ratios, and convert them to fractions.

    a. 20:120   c. 2.4:17.2

    b. 13:52   d. 2:12,000

2. At Homecoming, there are 45,000 fans rooting for the home team and 5,000 fans supporting the visiting team. What is the ratio Home Fans:Visiting Fans? What percentage of the stands are filled with visiting fans?

3. Your ecology class counted a total of 74 reptiles and amphibians in a one-square mile area near a pond. If 22% of the animals were reptiles, how many amphibians were counted?

4. Four out of five dentists agree that toothpaste is over-priced (I made that up; can you tell?) At a convention of 632 dentists, how many dentists would that be? What percentage think that toothpaste is *not* over-priced?

5. Your new job pays $37,500 per year. You work 50 forty-hour weeks in a year. What is your salary by the week? What is it by the hour?

6. A certain soap claims to be 99 and $\frac{44}{100}$% pure. In a ton of soap (2,000 pounds), how much "impurities" are there?

7. When heated to 200°C, a 120 inch metal rod lengthens by $\frac{3}{8}$ inch. By what percentage does the rod elongate? By what part per thousand?

8. A local reservoir is polluted with PCB's at a concentration of 20 ppm by volume. The reservoir holds 10 million gallons of water. What volume of PCB's are present?

Powers and Roots

The world of science is riddled with shorthand notation. This is not done to confuse outsiders, but rather to streamline discussions of complex ideas. The math that scientists use is likewise often streamlined to make manipulations of numbers and equations easier to fathom. Some mathematical descriptions fit this use of simplified notation. The next three chapters cover important examples: **powers (exponents)**, **roots**, **logarithms**, and **scientific notation**.

The use of **powers** arises naturally from the need to express a series of multiplications. For example, the number **64** is the result of multiplying the number **2** times itself six times:

$$2 \times 2 \times 2 \times 2 \times 2 \times 2 = 64$$

It is certainly easier to express this operation using the conventional notation 2^6. This shorthand style reads "two raised to the sixth power." The number **6** is the **exponent** which instructs you to multiply **2** times itself six times. So one may write:

$$2^6 = 64 \qquad [2 \ x^y \ 6 =]$$

which is more concise than the original equation above.

The series of **2** raised to greater and greater values of the exponent is referred to as the "powers of two." Two raised to the zero power, 2^0, is defined equal to **1**. In fact, *any* number raised to the power zero is equal to **1**. The series thus becomes:

$$2^0 = 1, \ 2^1 = 2, \ 2^2 = 4, \ 2^3 = 8, \ 2^4 = 16, \ 2^5 = 32 \ ...$$

The powers of two are very important in the science of computing. Computers "think" in 0's and 1's—the only digits necessary in base 2.

A very important series, one that is used throughout science, is the powers of ten:

$$10^0 = 1, 10^1 = 10, 10^2 = 100, 10^3 = 1000, 10^4 = 10{,}000 \ldots$$

The "milestones" represented by each power of ten are called **decades**, or "**orders of magnitude**." For instance, if two experiments yield data that differ by two orders of magnitude, then they are a factor of 100 off from one another (back to the drawing board!). Our entire number system is based on these powers of ten: the digits in a number represent ones, tens, hundreds, thousands, and so on. You will learn in Chapter 6 about **scientific notation**, an important representation of numbers using powers of ten.

As with most mathematical operations, **powers** have an inverse operation. The number **3** raised to the second power ("three squared") equals **9**. The inverse operation is to take the number **9** and ask "the square of what number will equal 9?" To answer this question you need to take the **second root** or the **square root** of **9**. You can represent this algebraically using the equation:

$$x^2 = 9$$

You must solve for the value of **x** using the square root operation. The symbol $\sqrt{}$ is used to represent taking the root of a number, hence $\sqrt{9} = 3$. For the second root (square root), the index **2** is not used: $\sqrt{9} = \sqrt[2]{9}$. Note that $(-3)^2 = 9$ as well, so $\sqrt{9}$ actually has two values, -3 and +3. Your calculator will only report the positive root.

Since exponents can have values larger (and smaller, as you will see below) than two, you must also be able to take roots larger than two. And such is the case. Referring to our example from above, the sixth root of **64** is equal to **2**:

$$\sqrt[6]{64} = 2 \qquad [64 \; x^{1/y} \; 6 =]$$

Thus, taking the roots of numbers is the opposite operation of raising numbers to exponents.

What about other exponents? If an exponent can be equal to zero, can it be negative? Of course—negative whole number exponents yield values less than **1**. The negative sign instructs you to divide the number into **1**:

$$10^{-1} = \frac{1}{10}$$

Note that 10^{-1} is the same as $1/(10^1)$, the reciprocal of 10^1. Numbers with negative exponents are thus the reciprocals of their positive exponent counterparts, for example $4^{-5} = 1/4^5$ and so on.

Can you have exponents that are not **integers**? Again the answer is "yes." The simplest examples are exponents that are fractions. Fractional exponents represent the inverse process—a number raised to a fraction actually indicates that a root should be determined. Using our familiar example from above:

$$9^{1/2} = 3$$

In other words, the exponent **1/2** instructs us to take the **second root** of **9**. So $9^{1/2}$ is another way of writing $\sqrt{9}$. Fractional exponents equal the roots: $64^{1/6} = \sqrt[6]{64} = 2$. This will become clearer when multiplication and division of exponential numbers is discussed below.

If exponents can be fractions, then that implies that they can have *any* decimal value. This is absolutely true, and is easily checked with your calculator. Try one, say, **5.03$^{-1.23}$**: [5.03 x^y 1.23 $\pm$ =]. It worked, didn't it? I got 0.137. To make the negative exponent you must enter the exponent *first* and then key to change the sign ($\pm$). You could not do that example by hand, although calculating powers by hand can be done for whole number exponents. Calculating roots is much more difficult, so you will rely on your calculator for these operations (you will naturally commit some perfect squares and cubes to memory through exposure and practice; quick, what is the cube root of 8?). You will use your calculator to take roots and most powers. Here are some examples:

$9^{1/9} = 1.277$	[9 $x^{1/y}$ 9 =]
$7^{-2} = 0.0204$	[7 x^y 2 $\pm$ =]
$e^3 = 20.086$	[3 e^x =]
$2^{-1/\pi} = 0.802$	[2 $x^{1/y}$ π $\pm$ =]

Multiplication and division of numbers with exponents can often be done by hand, even though you will probably still rely on your calculator to confirm your answer. This is a waste of time since you are actually doing the calculation twice! Instead, learn to recognize the simple calculations that can be done quickly by hand (or more correctly, in your brain), so that you skip writing steps and solve the problem faster). Let's first look at the mechanism for multiplying and dividing these numbers.

Multiplication of exponential numbers can be done quickly when the number being raised to the exponents (called the "base") is the same. For example $2^3 \times 2^6$ (here the base is **2**). The answer is obtained by *adding* the exponents (3+6=9) to obtain the new exponent:

$$(2\times2\times2)\times(2\times2\times2\times2\times2\times2)=2^9$$

$$2^3 \times 2^6 = 2^9$$

Thus the result is **512**. How about when exponents are negative? Try this one: **$10^{-3} \times 10^3$**.

The answer is 1. Add the exponents: **-3 + 3** is zero, and **10⁰** is one. A few more examples:

$$3^{-4} \times 3^6 = 3^{(-4+6)} = 3^2 = 9$$

$$2^{-2} \times 2^{-3} = 2^{(-2+(-3))} = 2^{-5} = 1/2^5 = 1/32$$

Division of numbers with exponents is similar to multiplication. Instead of adding exponents, you *subtract* them. For instance, the correct exponent derived from $2^3 \div 2^2$ is 1:

$$2^3 \div 2^2 = 2^{(3-2)} = 2^1 = 2$$

You subtract the second exponent from the first. Actually, you may find it easier to convert a division into a multiplication. Remember that the negative exponents represent reciprocals, like $1/(5^2) = 5^{-2}$. This means that any division can be converted to a multiplication by changing the sign on the exponent of the number in the denominator:

$$3^2 \div 3^{-4} = 3^2 \times 3^4 = 3^{(2+4)} = 3^6$$

Note that for the second number you just change the sign of the exponent (or multiply by **-1**, if you prefer) when you change the division to multiplication.

Does this trick make any difference? It might if you never can remember which exponent to subtract from which when doing a division. By converting the division to a multiplication you are left with the less complicated addition of exponents. Granted, you are using extra steps—and possibly time—to do the conversion. But if you have trouble organizing the manipulations in your mind, use the approach that gives you the right answer reliably.

Chapter 4 Summary

- **Powers** and **roots** are shorthand notation for complex multiplication (powers) and division (roots).

- **Exponents** can be any number—positive, negative, integer, or real.

- Negative exponents represent reciprocals: $10^{-5} = 1/10^5$ and $1/2^{-3} = 2^3$.

- Fractional exponents represent taking roots: $8^{1/3} = \sqrt[3]{8}$ and $32^{0.25} = \sqrt[4]{32}$.

- To multiply exponential numbers with the same base, add the exponents: $3^{2.5} \times 3^{0.5} = 3^3$ and $10^{-3} \times 10^{-5} = 10^{-8}$. When multiplying numbers with different bases, you must evaluate the exponential numbers separately before performing the multiplication.

- When dividing exponential numbers, subtract the exponent of the divisor (the denominator) from the exponent of the dividend (the numerator):

$$5^3 \div 5^{16} = 5^3/5^{16} = 5^{-13} \text{ and } 10^3 \div 10^{-6} = 10^3 \times 10^6 = 10^9.$$

Practice Exercises Chapter 4

1. Convert these numbers to exponential numbers:

   a. 100,000 (base 10)   c. 0.0001 (base 10)
   b. 512 (base 2)   d. 625 (base 5)

2. Evaluate these numbers without and then with your calculator:

   a. 7^4 d. 2^{-2}
   b. $(1.5)^3$ e. $\sqrt[3]{125}$
   c. 10^7 f. $(121)^{\frac{1}{2}}$

3. Determine these values with your calculator:

   a. $(2.2)^5$ d. $\sqrt[3]{\pi/2}$
   b. $(5.3)^{-2}$ e. $(4.21)^{-0.23}$
   c. $8^{-\frac{1}{3}}$ f. $e^{-\pi}$

4. Perform these calculations:

   a. $2^3 + 2^4$ c. $\sqrt{\pi^3 + \pi}$
   b. $10^3 - 10^2$ d. $\dfrac{10 - 10^2}{10 + 10^2}$

5. Perform these calculations:

   a. $(10^3)(10^{-2})$ c. $(2^{-\pi})(2^{-1})$
   b. $\dfrac{10^{-2}}{10^4}$ d. $(10^5)(10^{-\frac{1}{5}})$

Logarithms

Logarithms seem to give students headaches. Maybe this chapter will be the medicine that you need. Stated simply a logarithm is a **function**. That is, you apply it to a number to get another number in return. In scientific applications you will usually see two types of logarithms: the **common log** in base 10, and the **natural log** in base **e**. There is a way to convert between these two as you will see below. Logarithms can be useful in dealing with extremely large and small numbers—particularly when you must deal with them together. The "logarithmic scale" compresses orders of magnitude into segments that are equally separated.

The logarithm acts upon a number and returns a value that indicates the **exponent** of the number as it is expressed in base 10 for common logs or base **e** for natural logs. For example, suppose you had to develop a shorthand way of expressing numbers that could have the values 10^0, 10^1, 10^2, 10^3, 10^4, 10^5, or 10^6. An easy way to identify each number would be by the exponents 0, 1, 2, 3, 4, 5, or 6. You have just invented the logarithm (aren't you proud of yourself?).

Let's write your new function in a way that indicates what you did with the numbers above. Since you were using numbers in base 10 you need to indicate this fact. We can use an equivalence to express what the logarithm does:

$$\log_{10}(10^3) = 3$$

That's all there is to it. Your new function was applied to a number written in base 10 (10^3) and it returned the exponent 3. You can leave out the base 10 designation (the little subscript 10), since by convention the function **log** is assumed to be **log$_{10}$**. The natural logarithm would be **log$_e$** ("log to the base **e**"), but it has the special symbol **ln**. So try out your new function in base **e**:

$$\ln(e^{-5}) = ?$$

Right, the natural log of e^{-5} is just the exponent, -5 (note that the logarithm of a number with a negative exponent is a negative number).

Usually you will take the logarithm of a number that is somewhere between the decades in base 10. The examples above were easy, but not encountered that often. For instance what is the log of 70, log(70)? The number 70 lies between the number 10 (10^1) and 100 (10^2), so you expect that the log of 70 should be between **1**, log(10), and **2**, log(100). You need your calculator to evaluate the log(70) exactly:

$$\log(70) = 1.845 \quad [\ \mathbf{70\ log}\ ]$$

Note that to take the log you only enter the number (70 in this case) and press the key on your calculator that is labeled **log**. The result will be displayed without pressing the equals key. The same method is used to take natural logs, using the key labeled **ln**.

While logarithms appear to be straightforward (They do, don't they?—if not, go through it again. Take an aspirin first.), more students stumble using the inverse function, the **antilogarithm**. The antilogarithm is written using the shorthand notation **antilog**. Taking the antilog reverses the effect of taking the log, so if you take the antilog of the log of a number, you get that number back. In other words, if $y = 10^x$ then $y = $ antilog(x).

Let's start with a familiar log equation from above to show behavior of the antilog: $\log(10^3) = 3$. Apply the antilog to both sides of this equation,

$$\text{antilog}[\log(10^3)] = \text{antilog}(3)$$

Remember that the antilog of a log is just the number contained in the log brackets:

$$\text{antilog}[\log(10^3)] = 10^3$$

$$10^3 = \text{antilog}(3)$$

From here we can make a connection. The antilog of a number is obtained by making that number the *exponent*. Try another:

$$\text{antilog}(-1.33) = 10^{-1.33} = 0.0468$$

Antilogs present a challenge because often there is no "antilog" key on a calculator. There are a few ways to do the example above, depending on the capabilities of your calculator. If you have an inverse key, **INV**, then the keystrokes would be: **[1.33 ± INV LOG]**. Without the inverse function, you can use the **10^x** key: **[1.33 ± 10^x]**. If all else fails, use the **x^y** key: **[10 x^y 1.33 ± =]**. Following this model, you can also do the inverse of natural logs. For instance, the inverse natural log of -1.33 would be: [**1.33 ± e^x**] or [**2.71828 x^y 1.33 ± =**]. Note that when using the last method you need the value for **e.**

Sometimes you may need to convert between logarithms in base 10 and natural logarithms. Sometimes you will learn about a scientific relationship that was originally derived using the natural logarithm, but converted to an equation that uses the common logarithm. This is usually done to help students, since calculations in base 10 are easier to handle on lower priced calculators (it's not really necessary these days). A simple equation relates these two logarithms:

$$\ln(x) = 2.303 \log(x)$$

The conversion constant, 2.303, is an estimate of the full constant which is equal to ln(10). You can generate the exact constant with your calculator when you need it.

There are some manipulations of logarithms that you may come across. These relations will be useful when it is necessary to rearrange an equation that contains a logarithm:

1. **log(a^b) = b•log(a)** $\log(6^2) = 2 \log(6)$ and $\log(2^6) = 6 \log(2)$

2. **log(a) + log(b) = log(a•b)** $\log(6) + \log(2) = \log(6•2) = \log(12)$

3. **log(a) - log(b) = log(a/b)** $\log(6) - \log(2) = \log(6/2) = \log(3)$

You can use these same relationships with natural logs as well (i.e., $\ln(a^b)$ = bln(a), etc.). Note that log(a + b) is **NOT** equal to log(a) + log(b), and log(a)/log(b) is **NOT** equal to log(a/b)!

Chapter 5 Summary

- **Logarithms** are functions that yield the **exponent** from a number expressed in powers of 10 (common logs) or powers of **e** (natural logs): $\log(10^{-2}) = -2$, $\ln(e^6) = 6$.

- You will use your calculator to evaluate logarithms since most numbers will not correspond to 10 or **e** raised to a power. Use the **log** key for common logs and the **ln** key for natural logs.

- Antilogarithms are the inverse of logarithms: **antilog(x) = 10^x**.

- Common logs are related to natural logs by the equation:

$$\ln(x) = 2.303 \log(x).$$

Practice Exercises Chapter 5

1. Evaluate these logarithms:

a. $\log(1000)$ d. $\log(0.23)$

b. $\log(0.00001)$ e. $\ln(e^3)$

c. $\log(8)$ f. $\ln(2)$

2. Perform these antilogarithms:

a. antilog (-4) d. $e^{\ln(2)}$

b. 10^{-2} e. e^{-1}

c. $10^{\log(3)}$ f. $e^{2\ln(3)}$

3. Evaluate the following:

a. $\log(2)+\log(4)$ d. $\log(4)-\log(2)$

b. $\log(3^2)$ e. $\log(10\cdot100)$

c. $\ln(10^{-2})$ f. $\log(4+3)$

4. Convert these log expressions to natural logs:

a. $\log(2)$

b. $\log(e)$

c. $\log(100)$

CHAPTER 6

Scientific Notation

In your study of science, you will be confronted very early by the fact that our universe is a mixture of the very big and the very small. Consider for example the differences in the mass of the earth and the mass of one molecule of oxygen. The best current values are:

mass of earth = 5,980,000,000,000,000,000,000,000 kg

mass of oxygen = 0.0000000000000000000000000531 kg

(kg are kilograms, a unit of mass). It is not convenient at all to write out these long numbers! If it was part of your job to deal with numbers like these that were very large or very small you would soon throw up your hands in disgust—unless you turned to a shorthand notation to help. You can relax. This shorthand exists in the form of **scientific notation** (also called **exponential notation**).

Let's think of a number that's a little more exciting than the mass of the earth. What if you graduate from college and are rewarded with a gift from your rich uncle? How about a BMW 540i, list price $48,950? There are other ways to represent this number that will introduce the use of exponential notation. First, write the number after factoring out ten: 4895.0×10 (notice that when you factor out the ten you move the decimal one place to the *left*). Keep factoring out tens until you are left with a number with only one digit to the left of the decimal:

$$48,950 = 4895.0 \times 10$$
$$= 489.50 \times 10 \times 10$$
$$= 48.950 \times 10 \times 10 \times 10$$
$$= 4.8950 \times 10 \times 10 \times 10 \times 10$$

Now collect all the tens together, using the skill you developed with exponents in Chapter 4. Since four tens are multiplied together, this last number can be expressed as **4.8950**

× 10^4. There, you just did scientific notation. In this case it doesn't seem warranted because you are familiar with dollar amounts in the $50,000 range (or at least you *intend* to become accustomed to them after graduation!). By using this method the mass of the earth becomes 5.98×10^{24} kg, which is much nicer than the long version above.

Now try this with the mass of the oxygen molecule (mass of oxygen = 0.0000000000000000000000000531 kg). This time you won't factor out tens like you did in the BMW example above. In that example you moved the decimal to the left one place for each factor of ten. If you are trying to represent a number which is much less that one, you will do the opposite. You will want to move the decimal places to the right in order to write a very small number in exponential notation. Look at the number above, and count the number of places you must move the decimal to the right to get the decimal just after the 5 (5.31). You should get 26. Each time you move the decimal to the right, it is like factoring out a tenth, 10^{-1}. Since you have 26 of these, multiplied together, that yields 10^{-26}. So your resulting number in scientific notation is 5.31×10^{-26}. The large, negative exponent should jump out at you—this is a *very* small number.

So there you have it, big numbers have large positive exponents and small numbers are represented with large negative exponents. What other types of manipulations are made with these new shorthand numbers? You will most likely rely on your calculator for standard operations with exponential numbers, but you should understand how to make these maneuvers yourself—for two reasons. First, you can do an awful lot of calculating **without** your calculator to save time (see also discussion in Chapter 13 on Making Estimates). Also, there may be the occasion where your calculator may not handle the desired calculation (Horror of Horrors! Well, after all, it *is* only a machine). For example, pull out your nerdulator and try this one:

$$4.87 \times 10^{56} \div 2.23 \times 10^{-45} \ [\textbf{4.87 exp 56 + 2.23 exp 45 ± = }]$$

So what happened? If your calculator is like most, you got the big **-E-** . That's the symbol used by most calculators to mean that the result of your request is beyond the memory abilities of the circuitry.

You must prepare for this unfortunate limitation of your calculator. You certainly **do not** want to discover this during a class quiz. You need one of the rules for manipulation of numbers expressed in scientific notation. Let's look again at the calculation from above:

$$\frac{4.87 \times 10^{56}}{2.23 \times 10^{-45}}$$

Use your new skills with exponents to vanquish this one. The result of this division involves the division 4.87/2.23 (= 2.18) times the proper exponent determined by dividing 10^{56} by 10^{-45}:

$$10^{56}/10^{-45} = 10^{[56-(-45)]} = 10^{(56+45)} = 10^{101}$$

So your answer should be 2.18×10^{101}, not **-E-** .

Truth be told, however, this is not the most common type of scientific notation mistake that you may make. Most mistakes happen when addition or subtraction is attempted. The rule to keep in mind is that to add or subtract numbers represented in scientific notation you must have *both numbers* represented using the *same power of ten*. (This is not a problem for your calculator, yet you should master it for future estimation problems and for determining the proper number of significant figures—the topic of the next chapter.) You wouldn't add dollars to pennies without first converting the dollars to pennies ($2 + 33 cents = 200 cents + 33 cents = 233 cents) or vice versa ($2 + 33 cents = $2 + $0.33 = $2.33). Consider the addition of 2.13×10^5 and 7.55×10^3. Choose the number with the smallest exponent and rewrite the other number using **that** power of ten: $2.13 \times 10^5 = 213. \times 10^3$ (note that the decimal was moved two places to the right). Now you can add the two numbers:

$$
\begin{array}{r}
213. \times 10^3 \\
+\ 7.55 \times 10^3 \\
\hline
220.55 \times 10^3
\end{array}
$$

Now you can write the sum as 220.55×10^3, or change it to 2.2055×10^5 (decimal moved two places to the left). But wait—the original numbers only had three digits each, and this result has five. Is this right? Tune in to the next chapter to see how to represent this answer properly.

Chapter 6 Summary

- Big numbers and small numbers are written more conveniently in a format that takes advantage of exponents: **scientific notation**.

- Normal rules of exponents apply to calculations involving multiplication or division.

- In order to add or subtract numbers in exponential notation, both numbers must be represented in the same power of ten.

- Note that numbers like 10^6 are entered in your calculator as [**1 exp 6**], not [**10 exp 6**]. Other examples are 10^{-30} [**1 exp 30 ±**] and -10^{-12} [**1 ± exp 12 ±**].

- Rules of significant figures (Chapter 7) must be applied to correctly represent answers to calculations.

Practice Exercises Chapter 6

1. Convert the following to scientific notation:

   a. 54,000.00   c. −0.00131
   b. 142.35   d. 0.00000004

2. Convert the following from scientific notation:

a. 1.6×10^{-4} c. 3.7542×10^{3}

b. 0.2×10^{3} d. 4.0×10^{6}

3. Perform these calculations, without and then with your calculator:

a. $\left(3.4 \times 10^{-4}\right)\left(1.7 \times 10^{3}\right)$ c. $\left(0.2 \times 10^{-4}\right)\left(0.4 \times 10^{-4}\right)$

b. $\dfrac{3.4 \times 10^{-4}}{1.7 \times 10^{3}}$ d. $\dfrac{5.0 \times 10^{6}}{-2.5 \times 10^{-5}}$

4. Perform these calculations, without and then with your calculator:

a. $\left(2.2 \times 10^{-2}\right)+\left(1.4 \times 10^{-2}\right)$ c. $\left(4.53 \times 10^{2}\right)-\left(0.2 \times 10^{3}\right)$

b. $\left(5.6 \times 10^{4}\right)-\left(2.1 \times 10^{4}\right)$ d. $\left(1.1 \times 10^{-2}\right)-\left(3.0 \times 10^{2}\right)$

Significant Figures

Thank you for tuning in to Chapter Seven. In our last episode (the end of Chapter Six), you performed an addition of two numbers expressed in scientific notation:

$$213. \times 10^3$$
$$+ \ 7.55 \times 10^3$$
$$220.55 \times 10^3$$

The problem that was mentioned earlier involves the number of digits used in representing the sum—each number to be added has three digits, yet the sum is written with five. Or is it?

One of the challenges in using numbers to describe scientific information is your "confidence" in the numbers. Clearly, when you count discreet items like people, you are quite confident in the number—there are 24 people in your Spanish class, not 24.2 . However, for things that you measure, you are always faced with the task of representing how well you can make the measurement (the "error"). Once measurements are made and calculations are performed, then you are responsible for reporting the result in a way that reflects how any uncertainty in measurement affects the calculated answer ("propagation of error").

The correct way to propagate error involves application of differential calculus (wait...calm down, take a deep breath and read the next sentence...). Luckily, for calculations performed in first year science courses it is adequate to determine error in numbers using a much simplified system. By learning a few rules for manipulating **significant figures**, you can report numbers with the confidence expected in simple calculations. You will identify the digits in numbers that are significant and use the rules to represent results of calculations.

Consider the measurement of mass. Maybe you will determine the mass of some frogs for your biology lab. Suppose you measure one frog and it has a mass of 350 grams (about

29

three quarters of a pound, there are 453.6 grams per pound). How will you enter the mass of the frog in your lab notebook? It depends on how precise your measurement is—how carefully you read the balance and the sensitivity of the balance. If you used a sensitive balance which measures to the nearest hundredth of a gram (0.01 g) you could report 350.00 g. This number implies that the two numbers after the decimal are important or **significant**. Your uncertainty is in the last digit that you report (± 0.01 g). This is the simplified method for reporting uncertainty in measurements—when you write a number you will assume that your uncertainty is in the last significant digit. Now count all of the digits in the number 350.00. There are five significant digits.

You just learned one of the rules of significant digits—zeros to the **right** of the decimal are significant (they let you know exactly where your uncertainty lies). The zero in the ones place is also significant since zeros between non-zero digits (like the 5) and the last significant digit on the right (the last zero) are significant. What about zeros to the left of significant digits? The number 0.00350 has only **three** significant digits—the 3, 5, and the last zero on the right. The leading zeros are place markers for the decimal, and are not themselves significant. You can see this easily if we convert the number 0.00350 into scientific notation: 3.50×10^{-3} . This is an unambiguous way of representing this number with the proper significant digits. That is one of the powers of scientific notation—you can use it to properly represent numbers to the right number of significant digits.

Let us return to your hefty frog. What if the balance only read to the nearest ten grams? Reporting the number as 350 grams then becomes incorrect. Remember that when you write a number it is assumed that your uncertainty is in the last digit on the right. But the number 350 implies that your error is in the ones place. This is clearly wrong if the balance could only weigh items to the nearest ten grams. Here again you are saved by using scientific notation. If your uncertainty is in the tens place, then the digit that represents the tens must be the last digit on the right: 3.5×10^2. As you can see, the 5 is the last significant digit and it implies that the measurement was good to the closest ten grams. You would report the mass of the frog with two significant digits (the 3 and the 5).

Before showing how to manipulate numbers using proper significant digits, we need to discuss the issue of rounding. Often when you make a calculation you will need to round the result to the proper number of significant figures. Your calculator is no help in this regard—it keeps all digits whether they are significant or not. It is up to you to round correctly so that the result you report has the correct uncertainty. The simplest rounding rules are as follows:

1. If the digit in question is less than 5 then drop it and all digits after it. For example, you round the number 37.649899 to 37.6 when expressing it with three significant digits.

2. If the digit in question is 5 or more, increase the digit immediately before it by one. For example, 37.6500 and 37.683 will both round to 37.7.

Don't be fooled by impostors—these are the only rules you need!

Now you are positioned to use significant figures to work for you by revealing the uncertainty of measurements in calculations. Let's turn first to the example from the beginning of Chapter 3 that has been troubling us so much:

$$213. \times 10^3$$
$$+\ 7.55 \times 10^3$$
$$220.55 \times 10^3$$

The first number is 2.13×10^5, which has three significant digits. In order to add this number to 7.55×10^3, it is first necessary to get **both** numbers into the *same power of ten*. This is the rule for addition and subtraction. Once in the same power of ten, the numbers can be added, but the answer must be reported to the proper number of significant figures (**sig figs** for short). If your first number had no digits to the right of the decimal, then you cannot justify an answer that does—to do so would improve your precision artificially! Therefore the answer must be reported with the same number of digits past the decimal as the **least** number of digits that appear in the numbers being added or subtracted. That makes the result 220.55×10^3 incorrect. The proper answer would have no digits past the decimal: 221. (notice the rounding). The number 221. is correct, but it looks a little funny (with the decimal and no digit following). It is preferable to rewrite the number as 2.21×10^2. One good reason not to leave it as 221. is that the decimal might be mistaken for a period at the end of a sentence, or debris left by frogs hopping on your lab notebook.

The rules for reporting the result of a multiplication or division are somewhat simpler. When you perform the operation on two numbers, the result must have the same number of significant digits as the number with the **least** total significant digits. Say you measure a box with dimensions 6.4 inches by 4.55 inches by 2.35 inches. Your calculator spits out:

$$6.4 \text{ in} \times 4.55 \text{ in} \times 2.35 \text{ in} = 68.432 \text{ in}^3$$

Clearly you can't use all five of those digits for the volume. Your biggest uncertainty is in the first measurement of 6.4 inches. That number has only two significant digits—and so must your result. So the correct volume of the box is $6.8 \times 10^1 \text{ in}^3$. Notice that 68 in^3 is correct, but since there is no decimal it is not exactly clear where the uncertainty lies.

One rule of significant figures that seems to get lost in the shuffle is how to treat numbers involving logarithms. The rule is very simple, but students seem to forget it if they are ever even exposed to it (many instructors are guilty as well!). A practical example involves calculating the pH of a solution. The pH scale is logarithmic, and gives a quick indication of whether a solution is acidic (pH < 7) or basic (pH > 7). You might notice pH values in description of health and beauty aids like shampoos. The pH is determined by taking the log of the hydrogen ion concentration (hydrogen ions, H^+, are responsible for acidity), and multiplying the result by negative one:

$$pH = -\log[H^+]$$

If the hydrogen ion concentration is 3.5×10^{-3} M (the unit of concentration is M, molarity), the pH is:

$$pH = -\log[3.5 \times 10^{-3}] = -(-2.455931956) = 2.46 \quad [\textbf{3.5 exp 3} \pm \textbf{log} \pm]$$

This is about the pH of some soft drinks. Notice that the pH is reported with the number of digits to the right of the decimal (two) **equal** to the number of sig figs in the original number (two). All logs are reported the same way. Significant figures for antilog calculations are treated the same way. The antilog of -2.46 will have two significant digits since there are two digits to the right of the decimal:

$$antilog(-2.46) = 10^{-2.46} = 3.5 \times 10^{-3} \quad [\textbf{2.46} \pm \textbf{10}^x]$$

Chapter 7 Summary

- Assume that your uncertainty is in the **last** significant digit of a number.

- Zeros to the **right** of the decimal are significant: 5 sig figs for 2.0000 .

- Zeros between non-zero digits are significant: 4 sig figs for 30.04 .

- Leading zeros before the decimal are not significant: 2 sig figs for 0.0000040 .

- Use scientific notation to properly represent numbers to the right number of significant digits.

- Round numbers up when the digit is 5 or greater, down for less than 5: 246.552 rounds to 246.6 and 3449.2 rounds to 3.4×10^3.

- When adding or subtracting, it is necessary for **both** numbers to be the *same power of ten*.

- The result from an addition or subtraction has the same number of digits past the decimal as the **least** number of digits that appear in the numbers being added or subtracted: 2.45677 + 9.1 = 11.6 .

- When you multiply or divide two numbers, the result must have the same number of significant digits as the number with the **least** total significant digits: $0.0056 \times 345.56 = 1.9$.

- When taking logarithms, the number of digits to the right of the decimal in the result **equals** the number of sig figs in the original number: log(350.0) = 2.5441.

Practice Exercises Chapter 7

1. How many significant figures are in these numbers?

 a. 20.043   c. 0.004300

 b. 3.630×10^4 d. 100.0010

2. Round these numbers to three significant digits:

   a. 2.74450   c. 0.13251

   b. 532.62   d. 4223.0

3. Express the answer using the correct number of significant figures:

   a. $(3.33)(6.4290)$

   b. $(7.00)(8.3)$

   c. $\dfrac{4.0 \times 10^{2}}{0.2}$

   d. $(-0.05009)(1.2 \times 10^{8})$

   e. $78.834 - 78.8$

   f. $\dfrac{0.03303}{454.34}$

   g. $\ln(321.2)$

   h. $2.70 + \log(0.0020)$

Equations

If numbers are words in the language of science, then equations are the sentences. Sorry, that was a little hokey (but true). Equations express the relationships that are valuable in quantitative descriptions of the natural world. They allow us to state explicitly how certain parameters are linked, and use these links to predict behavior of systems as small as an atom and as large as the universe. Good equations make for a good scientific conversation.

Sentences are fairly predictable, they contain subjects, verbs, and objects. Equations also have a structure that you can always expect to see. From the simplest viewpoint, equations have a left side, a right side, and an equals sign between the two (in some cases, to be discussed later, the equals sign is replaced by an **inequality**). The equals sign is the heart of the equation. The left side and the right side are the same (equal!), even though they are expressed using different numbers or symbols. Let's explore a simple example to introduce these components.

Suppose you have a part-time job cleaning mouse cages for a genetics project in the Biology department, hauling in a steady $5.30 per hour. You need $250 for a new CD player. How many hours do you need to sling mouse mess to come up with the cash? The $250 is a **constant**, it is a number that *does not change in value*. Your hourly rate is also a constant. The only thing that changes is the number of hours that you work—it is called a **variable**. You can write down the relationship in an equation:

$$\$250 = (\$5.30 \text{ per hour}) \times (\text{hours worked})$$

To get a more compact equation you can replace "hours worked" with an algebraic variable, **x**, for instance. The equation becomes (dropping the units of dollars):

$$250 = 5.30\,\mathbf{x}$$

The equation above is the simplest type, containing only one variable and two constants. You can make this equation more general by assigning a variable to your total salary. The following equation describes your total salary (**S**) as a function of your hourly rate (5.30) and the number of hours worked (**x**):

$$S = 5.30 \, x$$

Here **S** takes the place of your total salary, which depends on your hourly rate (a constant in this example) and the number of hours that you work (a variable). Using this equation you can determine how much you will earn for any number of hours worked. By replacing **x** with the number of hours, you can calculate a total salary.

The key to using equations is rearranging them to give the value that you seek. Rearranging equations is a skill that you need to develop if you intend to be able to solve problems in science. It is simpler than it appears, and with some practice you will be able to rearrange equations quickly and see relationships that are important to your understanding of a subject. Consider the bottom line: you can only do to one side of an equation *what you do to the other*. Thus

$$2 \, S = 2(5.30) \, x$$

since you multiplied both sides by **2**. Let us rearrange this equation, isolating **x** on the left side. Normally you isolate the variable you seek to determine on the left side of the equation. The first step is to rewrite the equation by simply exchanging the left and right sides:

$$2(5.30) \, x = 2 \, S$$

Next the numbers modifying **x**, the **coefficients**, can be eliminated by multiplying both sides of the equation by the reciprocals:

$$\frac{1}{2(5.30)} \times 2(5.30) x = 2 S \times \frac{1}{2(5.30)}$$

The right side can be rewritten in a simpler format:

$$\frac{1}{2(5.30)} \times 2(5.30) x = \frac{2S}{2(5.30)}$$

The resulting equation isolates **x** on the left:

$$x = \frac{2S}{2(5.30)}$$

since the coefficients of **x** multiplied by their reciprocals is equal to **one**. The factor of **2** that we introduced earlier can be eliminated, since it appears on the top and bottom of the fraction. The final relation becomes:

$$x = \frac{S}{5.30}$$

This equation shows that to find the number of hours needed (**x**) you simply divide your total desired salary (**S**) by your hourly rate (5.30). So to get that CD player you need to work:

$$x = \frac{250}{5.30} = 47.2 \text{ hours}$$

Most equations will require manipulation for you to solve them when doing problems. Once you have identified the correct equation (a job in itself) you must rearrange it properly so that you can input numbers and perform a calculation of the answer. In the example above, we wrote an equation using the information in the problem—we determined a relationship between the variables and constants. In practice you will be presented with information (data) that you will use in an equation that you have learned to apply in certain situations. By moving variables around you will rearrange the equation so that you can determine an answer to the problem. Remember, you can only solve for *one variable* with a single equation—you must have a value for **all** other variables and constants. If you have an equation with two unknown variables, you will need another equation that relates the two. We'll look at this again in Chapter 12, Solving Problems. Right now let's look at other examples of equations and how to rearrange them.

The rearrangement of the salary equation was done in great detail, step by step. As you use equations you will make manipulations faster by moving variables and constants in smoother, multiple steps. Let's redo the example, starting with:

$$2\,S = 2(5.30)\,x$$

Since the **2** appears on both sides, it can be factored out (eliminated from both sides of the equation). To get the **x** alone we need to divide both sides by 5.30, leaving:

$$\frac{S}{5.30} = x$$

which is quickly reversed to yield our final equation.

$$x = \frac{S}{5.30}$$

With practice you will be able to rearrange equations in larger steps without writing intermediate steps down on paper. Consider Coulomb's Law, which describes the force F between two particles of charge q and Q separated by distance r:

$$F = k\frac{qQ}{r^2}$$

where k is a constant. Suppose you needed to solve for **r**. You should recognize that to get r alone you must first get r^2 alone. This is accomplished by multiplying both sides by r^2 to yield:

$$r^2 F = kqQ$$

and dividing both sides by F:

$$r^2 = \frac{kqQ}{F}$$

To obtain this equation, all you did was to switch places between the variables r^2 and F in the original equation. This illustrates a common manipulation in equation rearrangement when the left and right sides consist of a single fraction. When you move a variable from the *denominator* of a fraction on one side, it takes its place in the *numerator* of the other side. The reverse is also true, so the F on the left (*numerator* of left side) moves over to the *bottom* of the fraction on the right. Now the value of r can be determined by taking the square root of **both sides** of the equation:

$$\sqrt{r^2} = \sqrt{\frac{kqQ}{F}}$$

so that:

$$r = \sqrt{\frac{kqQ}{F}}$$

The previous examples involve variables and constants that are multiplied or divided only. Often equations involve terms that are added or subtracted. Remember, when rearranging an equation you must do the same mathematical manipulation to **each** side. So if you subtract a number from the left side you must also subtract the same number from the right side. When you move a term from one side to the other, its value becomes the negative of its previous value. For example, the following equation relates the Fahrenheit and Celsius temperature scales:

$$T_F = (\tfrac{9^\circ F}{5^\circ C}) T_C + 32^\circ F$$

To solve for the temperature in Celsius, T_C, first subtract 32°F from each side,

$$T_F - 32^\circ F = (\tfrac{9^\circ F}{5^\circ C}) T_C$$

(note how the 32°F moves to the left side and becomes **negative**) then multiply each side by the reciprocal of the conversion constant:

$$(\tfrac{5^\circ C}{9^\circ F})(T_F - 32^\circ F) = (\tfrac{5^\circ C}{9^\circ F})(\tfrac{9^\circ F}{5^\circ C}) T_C = T_C$$

and reverse the equation to isolate T_C:

$$T_C = (\tfrac{5^\circ C}{9^\circ F})(T_F - 32^\circ F)$$

So much for the easy stuff. Let's look at an equation that combines the challenges of multiplication/division with addition/subtraction. For a sample of gas, the van der Waals equation of state relates pressure (P), volume (V), number of particles (n, the number of moles), and temperature (T). There are two van der Waals constants, a and b, which are different for different gases. This equation is an improvement on the "ideal" gas law (which contains the gas constant, R),

$$PV = nRT$$

where the variables P and V have been expanded using the van der Waals constants:

$$(P + \frac{an^2}{V^2})(V - nb) = nRT$$

To rearrange this equation to solve for P, first divide both sides by (V - nb)—in other words, move the (V - nb) factor into the denominator of the right side:

$$(P + \frac{an^2}{V^2}) = \frac{nRT}{(V - nb)}$$

To isolate P, you must move the term containing a by subtracting $\dfrac{an^2}{V^2}$ from both sides:

$$P = \frac{nRT}{(V-nb)} - \frac{an^2}{V^2}$$

There, that wasn't so bad, was it? Now try to solve for the constant b :

$$\left(P + \frac{an^2}{V^2}\right)(V-nb) = nRT \qquad \text{original equation}$$

$$V - nb = \frac{nRT}{\left(P + \frac{an^2}{V^2}\right)} \qquad \text{isolate V-nb term}$$

$$-nb = \frac{nRT}{\left(P + \frac{an^2}{V^2}\right)} - V \qquad \text{move V to right side}$$

$$b = \frac{V}{n} - \frac{RT}{\left(P + \frac{an^2}{V^2}\right)} \qquad \text{multiply each side by } \frac{-1}{n}$$

Note that when you multiplied the right side by the negative number, the positions of the terms reversed to keep the positive term first. In other words:

$$b = \frac{-1}{n}\left[\frac{nRT}{\left(P + \frac{an^2}{V^2}\right)} - V\right] = \frac{-RT}{\left(P + \frac{an^2}{V^2}\right)} + \frac{V}{n} = \frac{V}{n} - \frac{RT}{\left(P + \frac{an^2}{V^2}\right)}$$

When you multiply each side of an equation by a negative number, you will *reverse the order* of terms in a subtraction. Consider the general case

$$a - b = c - d$$

where each side of the equation is multiplied by -1:

$$-(a-b) = -(c-d)$$

$$-a+b = -c+d$$

which becomes:

$$b - a = d - c$$

Special Equations

Some special formats for equations will come up often in problems that you solve. One such equation is the equation for a line:

$$y = mx + b$$

which we will cover in detail when we discuss graphing in Chapter 10. Another common equation involves a single variable that occurs in terms that are squared. For example, can you solve the following equation in **x**?

$$(x-2)^2 = 0$$

The solution is 2, since only 2 – 2 will give zero. This equation can be written another way by multiplying out the terms:

$$(x-2)(x-2) = 0$$

so that:

$$x^2 - 4x + 4 = 0$$

This form of the equation is called a **Quadratic Equation**, and it involves terms of x which are squared. The general form for a Quadratic Equation is:

$$ax^2 + bx + c = 0$$

There is a term in the square of **x** with coefficient a, a term in **x** with coefficient b, and a constant c. The equation is manipulated so that the right side of the equation is equal to **zero**. In other words, the equation:

$$3 - 2x = 3x^2$$

should be rearranged to the form:

$$3x^2 + 2x - 3 = 0$$

before you attempt to solve for the variable **x**.

Let's review how to multiply out terms like those mentioned above. For example, multiply the terms in parentheses:

$$(2x-1)(x-2)=0$$

The result is the following sum:

$$(2x)(x)+(2x)(-2)+(-1)(x)+(-1)(-2)=0$$

Note how every term was multiplied times the others. Note also that the terms are written in a specific order. This order comes from the FOIL technique, which stands for **First, Outer, Inner, Last**. Sum the product of the first terms, followed by the product of the outer terms, inner terms, and last terms. Simplify this sum to:

$$2x^2+(-4x)+(-x)+(2)=0$$

$$2x^2-5x+2=0$$

By inspecting this last equation, it is not obvious what values x must assume to satisfy the equation. You can use the "factored" version (the original version) of the equation to solve for **x**:

$$(2x-1)=0 \ \ \text{or} \ \ (x-2)=0$$

since only one factor need be equal to zero to satisfy the equation.

The two values of **x** that will satisfy the equation are:

$$x=\frac{1}{2} \ \text{or} \ x=2$$

Note that there are **two** solutions to a Quadratic Equation. It would be difficult to determine those solutions given the expanded equation $2x^2-5x+2=0$. However, it may also take you some time to factor a Quadratic Equation into terms that you can solve. Luckily there is a fast way to determine the solutions to a Quadratic Equation.

Remember that all Quadratic Equations can be written in the form:

$$ax^2+bx+c=0$$

Use the coefficients a, b, and c to solve the equation using the **Quadratic Formula**:

$$x = \frac{-b \pm \sqrt{b^2 - 4ac}}{2a}$$

Note that there will be two values of **x**, since the $\pm$ symbol instructs you to do the calculation two ways—one adding the term in the square root symbol and the other by subtracting it. So for our example,

$$a = 2 \quad b = -5 \quad c = 2$$

therefore,

$$x = \frac{-(-5) \pm \sqrt{(-5)^2 - 4(2)(2)}}{2(2)}$$

will give the following solutions:

$$x = \frac{5 \pm \sqrt{25 - 16}}{4} = \frac{5 \pm \sqrt{9}}{4} = \frac{5 \pm 3}{4} = 2, \frac{1}{2}$$

The Quadratic Formula gave the same solutions that we determined from the factored form. Just remember to set your quadratic equation equal to **zero** so that you can use the Quadratic Formula.

Inequalities

Some problems will require that you not equate things, but rather you will set limits using inequalities. You might encounter the inequalities **less than**, **<**, **less than or equal to**, **≤**, **greater than**, **>**, or **greater than or equal to**, **≥**. You should treat equations that contain inequalities *the same* as equations with an equals sign, except in one very important way. When you multiply or divide your equation by a **negative** number, you must **reverse** the direction of the inequality (< becomes >, and > becomes <, for instance). A simple example will illustrate the point. The following inequality is true:

$$-2 < +5$$

If you were to multiply this equation by -3, you must reverse the inequality to keep the equation true:

$$(-3)(-2) ? (-3)(+5)$$

$$+6 > -15$$

This treatment of inequalities is the same for equations that contain variables as well. As an example, let's solve for **x** in the following equation:

$T - 6x \geq 12T^2 + 3$ original equation

$-6x \geq 12T^2 + 3 - T$ move **T** to the right side

$(-\frac{1}{6})(-6x) \leq (-\frac{1}{6})(12T^2 - T + 3)$ multiply by inverse of **-6**, reverse the inequality!

$x \leq -2T^2 + \frac{T}{6} - \frac{1}{2}$ final equation solved for **x**

Note that when you multiplied by $-\frac{1}{6}$ you had to reverse the direction of the inequality. It is critical that you remember to do this when working with inequalities.

Chapter 8 Summary

- Equations have a left side, a right side, and an equals sign or inequality between the two.

- Components of equations can be **variables** (things that change), their **coefficients** (numbers multiplied times variables), or **constants** (numbers that do not change).

- You can only do to one side of an equation *what you do to the other*.

- Only one variable can be solved using a single equation. Two equations are required to solve for two variables.

- For equations with a single fraction on each side, when you move a variable from the *denominator* of a fraction on one side, it takes its place in the *numerator* of the other side. The reverse is also true (*numerator* of one side goes to *denominator* of the other).

- When you multiply each side of an equation by a negative number, you will *reverse the order* of terms in a subtraction. The relation $a - b = c - d$ becomes $b - a = d - c$ when multiplied by –1.

- The equation for a line is $y = mx + b$.

- The general form for a Quadratic Equation is $ax^2 + bx + c = 0$. The right side must equal to **zero** to use the Quadratic Formula.

- The **Quadratic Formula** is solved using the relation:

$$x = \frac{-b \pm \sqrt{b^2 - 4ac}}{2a}$$

- Equations can contain the inequalities less than, <, less than or equal to, ≤, greater than, >, or greater than or equal to, ≥.

- When multiplying or dividing an inequality by a **negative** number, you must **reverse** the direction of the inequality.

Practice Exercises Chapter 8

1. Solve the following equations for x:

   a. $42 = 14x$

   c. $3x^2 = 75$

   b. $\dfrac{(x+2)}{(x-2)} = 2$

   d. $\dfrac{3x}{(x+1)} = 2$

2. Rearrange these equations to solve for y:

   a. $2x + 2y = x^2$

   d. $2x^2 - x - 1 = 4xy - 2y^2$

   b. $E = RT \ln\left(\dfrac{1}{y}\right)$

   e. $2\ln x + \ln y = 2$

   c. $2 - y = 3a - x$

   f. $\dfrac{y+1}{x+1} = 2$

3. Rearrange these equations to the form of a line ($y = mx + b$).

   a. $\dfrac{y+5}{x+1} = 3$

   c. $\dfrac{E}{h} = v - v_{\circ}$ $\left(v_{\circ} \text{ is a constant}\right)$

   b. $k = Ae^{-2x}$

   d. $2y = \dfrac{x^2 - 2x + 1}{x - 1}$

4. Solve these quadratic equations (use the quadratic formula when necessary):

   a. $4x - x^2 = 4$

   c. $x^2 + 4x + 3 = 0$

   b. $x = \dfrac{6}{x} - 1$

   d. $x(1+x) = 0.75$

5. Solve these inequalities for x:

   a. $y - x \geq 2$

   c. $\ln x > 2 - y$

   b. $y < x^2 + 2$

   d. $2 - \dfrac{1}{x} \leq y^2$

CHAPTER 9

Units

Units give value to numbers. They tell us what measurement has been made, and in what magnitude. Consider, for example, distance. The number "5" does not convey enough information about a distance. Is it 5 inches, or 5 miles? There is a big difference. What about the differences between **m**, **mm**, **cm**, and **km**? Prefixes of units can relate orders of magnitude. By knowing units, their prefixes, and how to use them, you will help solidify your understanding of numbers, equations, and the measurements used in science.

Units fall into one of two categories, the *fundamental* units and the *derived* units. The fundamental units include length, mass, time, and temperature. Derived units are combinations of fundamental units. For example, velocity is found from distance (length) divided by time, so velocity has units of length divided by time:

$$\text{velocity} = \frac{\text{length (meters)}}{\text{time (seconds)}} = \frac{\text{m}}{\text{s}}$$

Another unit of velocity would be miles per hour, again a length (miles) divided by time (hours). All derived units involve application of an equation using the proper fundamental units (or other derived units). That is the power of using units: if you manipulate units properly the answer you determine from a formula will have the right units. Always "carry" your units, they will guide you to the right answer. Once your computation is done, if the result has the wrong unit then it will raise a red flag. You can detect mistakes by analyzing why your calculated unit is in error.

Scientists use units of the International System of Units (Le Système International d'Unités, or **SI**). Table 9-1 lists **SI** base units used around the world. There are also standardized prefixes used with units (those listed in Table 9-1 and derived units as well). These prefixes represent the order of magnitude by which the unit is multiplied. For instance, 1000 grams (1000 g) is a kilogram (1 kg), since **kilo** is the prefix for 1000. These SI prefixes are listed in Table 9-2. The most commonly used prefixes are **nano**, **micro**, **milli**, **centi**, **kilo**, **mega**, and **giga**.

Table 9-1 SI Base Units

Quantity	Unit	Symbol
Length	meter	m
Mass	kilogram	kg
Time	second	s
Temperature	kelvin	K

Converting units is a skill that you must master in order to correctly solve problems. Conversions are of three types: conversions that involve orders of magnitude (1000 mm per meter), conversions that involve definitions (60 s per minute), and conversions between measurement systems (2.54 cm per inch). For this last type, instead of trying to remember the limitless number of conversion factors, just focus on two that will allow conversion of length and mass: 2.54 cm/in and 453.6 g/lb. If you can remember these two, and convert your numbers to cm/in or g/lb where appropriate, you will always be able to make the proper conversion. A conversion is really a ratio that has a value equal to *one* (you can always multiply by the number one). You multiply the number you want to convert by the conversion expressed with the proper combination of units.

The following example shows all three types of conversions. Note that in each step you cancel one unit by multiplying or dividing by the proper conversion. The result is the same number expressed using the new unit.

The earth is about 144 billion meters from the sun. How far is this in miles?

$$144 \times 10^9 \, m \left(\frac{100 \, cm}{1 \, m} \right) = 144 \times 10^{11} \, cm \qquad \text{[order of magnitude change]}$$

$$\frac{144 \times 10^{11} \, cm}{2.54 \, cm/in} = 5.67 \times 10^{12} \, in \qquad \text{[system conversion]}$$

$$\frac{5.67 \times 10^{12} \, in}{12 \, in/ft} = 4.72 \times 10^{11} \, ft \qquad \text{[inches to feet, should know]}$$

$$\frac{4.72 \times 10^{11} \, ft}{5280 \, ft/mi} = 89.5 \times 10^6 \, mi \qquad \text{[feet to miles, handy to know]}$$

So the earth is about ninety million miles from the sun. The strategy was to convert meters to centimeters so that you could use your centimeters-to-inches conversion factor. Once in inches, you can get to miles using the definitions of feet and miles. It took a few steps, but it is logical and easy to do, at least easier than remembering that there are 1609 meters per mile! You can combine the conversions into one big equation:

Table 9-2   SI Prefixes

Prefix	Symbol	Value
atto	a	10^{-18}
femto	f	10^{-15}
pico	p	10^{-12}
nano	**n**	$\mathbf{10^{-9}}$
micro	**μ**	$\mathbf{10^{-6}}$
milli	**m**	$\mathbf{10^{-3}}$
centi	**c**	$\mathbf{10^{-2}}$
deci	d	10^{-1}
deka	da	10^{1}
hecto	h	10^{2}
kilo	**k**	$\mathbf{10^{3}}$
mega	**M**	$\mathbf{10^{6}}$
giga	**G**	$\mathbf{10^{9}}$
tera	T	10^{12}
peta	P	10^{15}
exa	E	10^{18}

$$\frac{(144 \times 10^{9} \text{ m})(100 \text{ }^{cm}/_{m})}{(2.54 \text{ }^{cm}/_{in})(12 \text{ }^{in}/_{ft})(5280 \text{ }^{ft}/_{mi})} = 89.5 \times 10^{6} \text{ mi}$$

The advantage of using one conversion equation is that you can see more clearly how the various units cancel to yield a number in the desired unit. Units that appear both in the denominator and numerator will cancel. The drawback is that you have to be careful to avoid losing your place when making decisions about whether to put a conversion in the numerator or the denominator. Notice from the example above that the conversion factors are exact numbers and do not affect the multiplication and division rules of significant figures. The use of 12 in/ft does not make the final answer 9.0×10^{7} mi.

In the previous example, conversions were expressed in fractional form. Often units are written using an "in-line" format. Negative exponents are used for units that appear in the denominator of derived units. In this alternative format, 2.54 cm/in becomes 2.54 cm·in⁻¹. Force is measured in the derived unit *newtons*, N. Recall that force is equal to mass (kg) times acceleration ($m/_{s^2}$):

$$F = m(kg) \cdot a(m/_{s^2}) = (m \cdot a) \frac{kg \cdot m}{s^2}$$

Using the in-line format, newtons can be expressed as $kg \cdot m \cdot s^{-2}$. The raised periods represent multiplication of the fundamental units. Learn to recognize the in-line format by using it when you do practice exercises. The fractional format may be easier for you to use (it is for me), so you will need to convert freely between the two formats.

One pitfall to watch out for is the use of conversion factors that are raised to powers. Consider a 10 ft × 10 ft × 8.5 ft room with a volume of 850 ft³. When you convert this volume to cubic meters, you must also cube the conversion factors:

$$850 \, ft^3 \times \left(\frac{12 \, in}{1 \, ft}\right)^3 \times \left(\frac{2.54 \, cm}{1 \, in}\right)^3 \times \left(\frac{1 \, m}{100 \, cm}\right)^3 =$$

$$850 \, ft^3 \times \left(\frac{1728 \, in^3}{1 \, ft^3}\right) \times \left(\frac{16.4 \, cm^3}{1 \, in^3}\right) \times \left(\frac{1 \, m^3}{10^6 \, cm^3}\right) = 24 \, m^3$$

Chapter 9 Summary

- Units are important—they indicate what measurement has been made, and in what magnitude.

- Units fall into two categories, *fundamental* units and *derived* units.

- Always "carry" your units, they will guide you to the right answer. By canceling units as you use an equation, you will avoid mistakes.

- Learn the **SI** base units and standardized prefixes.

- Converting units is a skill that you must learn.

- A unit conversion is really a ratio that has a value equal to *one*.

- Commit the conversions 2.54 cm/in and 453.6 g/lb to memory.

- Combine conversions into one big equation for clarity.

- Cancel units that appear both in the numerator and denominator.

- Units are sometimes written in an "in-line" format using negative exponents for units that appear in the denominator.

- Don't forget to raise conversion factors to their powers correctly.

Practice Exercises Chapter 9 ─────────────────────────

1. Change these numbers to equivalents using unit prefixes (for example, 1000g = 1kg):

   a. $5,000,000   e. 8×10^{-3} W

   b. 2,000,000,000 bytes   f. 10^{-6} m

   c. 0.025m   g. 50×10^{3} tons

   d. 4×10^{-9} s   h. 1×10^{12} dactyls

2. Rewrite these numbers without using the unit prefix:

   a. 5.3 Mbar   c. 20mm

   b. 0.2 kJ   d. 1.1 ng

3. Perform these unit conversions:

   a. 1.0 yr to s   e. 0.35 lb to kg

   b. 5.29×10^{-11} m to pm   f. 1.2 μs to ns

   c. 500 yd to mi   g. 101.3 MHz to kHz

   d. 0.0025 mi to m   h. 100 g to lb

4. Change these units to the in-line format:

   a. $96485\dfrac{C}{mol}$ c. $2.998\times10^{8}\dfrac{m}{s}$

   b. $0.08206\dfrac{atm \cdot L}{mol \cdot K}$ d. $1\dfrac{kg}{m \cdot s^{2}}$

5. Change these units from in-line format to fractional format:

   a. $1J \cdot s^{-1}$ c. $8.314 J \cdot mol^{-1} \cdot K^{-1}$

   b. $9.81 m \cdot s^{-2}$ d. $1 kg \cdot m^{2} \cdot s^{-2}$

Graphing

Graphs are pictures that express mathematical relationships. You are probably already familiar with many types of graphs since they are appearing more often in the print media and on television. Even presidential candidates use graphs to try and sell their economic plans on TV. The most common graphs used in the media are the **bar chart** and the **pie chart**. Pie charts are circles divided into segments. Each segment represents a fraction of the whole contributed by a characteristic. In a recent poll, one thousand chemistry students were asked to indicate their gender for statistical purposes as part of a survey. The results are presented in Figure 10-1 in the form of a pie chart. What is your interpretation of this data?

The bar chart differs from the pie chart. In a pie chart the sum of all segments must equal 100%. The bar chart represents behavior of one parameter as a function of another. A common use of bar charts is to show how some characteristic varies with time. This graphical representation is often used to show the fluctuations in the Stock Market on the evening news. The length of the bar represents the Dow Jones Industrial Average (an overall gauge of stock value), and the position of the bars typically advances from left (early time) to right (later time) at regular intervals. Figure 10-2 shows such a chart for the

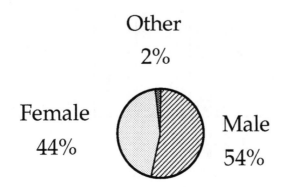

Figure 10-1 *Example of a pie chart.*

year 1987. A close-up of October (insert in Fig. 10-2) shows "Black Monday," October 19th, when the Dow Jones Industrial Average tumbled almost 500 points (data from *The Dow Jones Averages*© *1885-1990*, edited by Phyllis S. Pierce, Business One Irwin, Homewood, Ill., 1991). It is easy to understand the fluctuations in stock values at a glance by noting the variation in the height of the bars (this will become considerably more important to you after graduation when you are making major bucks).

Bar charts are useful when each value along the horizontal axis (the **x-axis**) can have only one amount associated with it. Often this is not the case. In addition, for a given set of information (data), the value of the parameter represented along the x-axis (the *independent variable*) may not necessarily occur in regular intervals. We can represent the measured parameter (the *dependent variable*) as a function of the independent variable in a more flexible format if we let the resulting value be a point on a two dimensional grid. A collection of points that represents this behavior is called a **scatter plot**. Points have **coordinates** that represent values of the independent variable (x) and the dependent variable (y). The dependent variable is plotted on a vertical axis (**y-axis**) that is positioned perpendicular to the x-axis. Each point has a position determined by two parameters, x and y, so the coordinates are written in the format **(x,y)**. Figure 10-3 shows the Stock Market data replotted in the scatter plot format, with lines joining points to aid the eye.

A scatter plot is usually just the beginning in treatment of scientific data. In constructing an experiment, you normally control one variable (time or temperature, for example) and make measurements of the dependent variable relative to it. This data becomes a

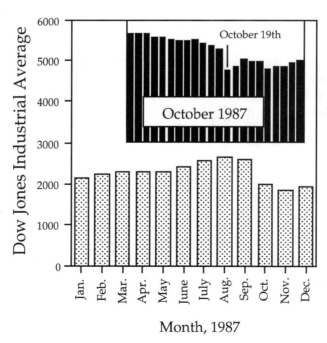

Figure 10-2   *Bar chart of the Dow Jones Industrial Average for the year 1987.*

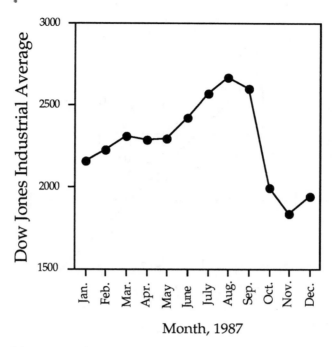

Figure 10-3   *Scatter plot of data presented in Fig. 10-2.*

collection of (x,y) pairs. Once these points are plotted, the result may follow a specific relationship. One of the most important relations is the equation of a line. The value of y behaves predictably as a function of x. The equation has a particular form:

$$y = mx + b$$

where m is the **slope** of the line and b is the **y-intercept**. The y-intercept is the value of y when x = 0 (where the line crosses the y-axis). The slope is the "tilt" of the line, determined by how much y values change with respect to x values. This change is the "rise over run," which means that you can determine the slope by comparing two points on the line, (x_1, y_1) and (x_2, y_2):

$$m = \frac{"rise"}{"run"} = \frac{y_2 - y_1}{x_2 - x_1} = \frac{\Delta y}{\Delta x}$$

The symbol Δ (delta) means "change in" and instructs you to take a difference between two values. Two points determine a line, and the equation of the line can be determined from *any* two points on the line. The traits of a line are summarized in Figure 10-4. The slope of the line in Figure 10-4 is negative. A line with m > 0 (positive) slopes the other way.

When treating experimental data that should follow a linear relationship, it is anticipated that the (x,y) pairs will fall on a line with a particular slope and y-intercept.

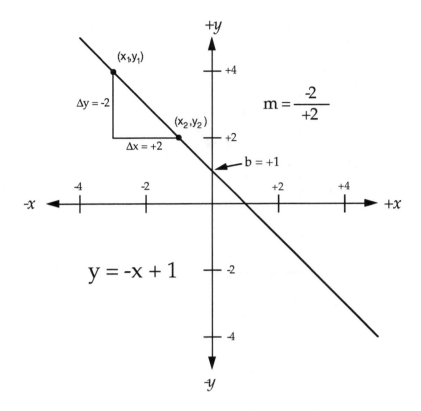

Figure 10-4 *A line with the equation y = -x + 1.*

Typically data is plotted to extract a value for the slope (and occasionally the intercept) because it represents a characteristic of the experimental system. Any equation that has the format y = mx + b can be plotted and the slope and intercept determined from any two chosen points. The relation above is used to determine the slope, and values of x and y for one point can be input with the slope to determine the intercept:

$$b = y - mx$$

Once the equation for a line is known, all points can be determined, particularly those that lie between data points (**interpolation**) and those that lie outside of the data range (**extrapolation**).

New values for y can be interpolated or extrapolated from the slope, intercept, and a given value of x by inputting all known information into the y = mx + b equation:

$$y_{new} = mx_{new} + b$$

Interpolation can also be performed for situations in which the data follows a curve, if the behavior between two closely spaced points can be approximated by a line. In this instance, the two data points are used to "define" a line of interpolation and the new point is determined in the normal fashion. This method is valuable only if the functional behavior is close to linear between the two known data points. For the same reason care must be exercised when extrapolating values beyond the range of collected data. If the line obtains curvature outside of the data range then an extrapolation will yield erroneous results. Only when a function is known not to deviate from linear behavior can extrapolations be performed successfully. Never stake your reputation on an extrapolation. There have been numerous examples of linear relationships that deviate at higher and lower values. In fact, these deviations themselves can become a field of study. Nonlinear effects are becoming increasingly important as we stretch our understanding of systems both ultra-large and ultra-small.

The ease of fitting data to a straight line can be exploited with equations that are not normally linear. When appropriate a relation can be converted from its original form to a linear format so that measurements can be plotted as (x,y) pairs and a slope can be determined. One common transformation of an equation is to take the logarithm (roots are also useful in this regard). The new form of the equation is set up in the linear fashion which parallels y = mx + b. One example is that of an equation that contains an exponential term:

$$k = Ae^{-E_a/RT}$$

This is the Arrhenius Equation that relates rates of chemical reactions to temperature. When the natural logarithm is applied to this equation it becomes:

$$\ln(k) = \frac{-E_a}{R}\left(\frac{1}{T}\right) + \ln(A)$$

Values of k are converted to $\ln(k)$ and plotted versus the reciprocal of temperature (1/T). A best fit line yields a slope of $m = \frac{-E_a}{R}$. Therefore a value for the activation energy (E_a) of a biochemical reaction, for example, can be determined graphically from a plot of the natural log of rate constant k as a function of reciprocal temperature.

Of course it is rare that a collection of data will form a perfect straight line. The experimenter must place a line through the data that represents the "best fit" linear relation. When plotted points fall very near a straight line this can be done "by eye" using a straight edge and good judgment. However, to be more exacting there is a well established method for determining the best fit straight line to a body of data. This technique is called the Least Squares Method. If you do all of your graphing using a computer or graphing calculator this method is but a few keystrokes away. It is valuable to understand how this method works (and where the name comes from). Figure 10-5 (a) shows the best fit straight line through a scatter plot of (x,y) data. Each data point is connected to this best line using a square with sides equal to y - y_{fit}. For the best fit line, the sum of areas for all of these squares will be at a minimum. Compare the best fit line in Figure 10-5 (a) with an alternate line in Figure 10-5 (b). For the best line the sum of the areas of the squares is smaller. Hence the name Least Squares.

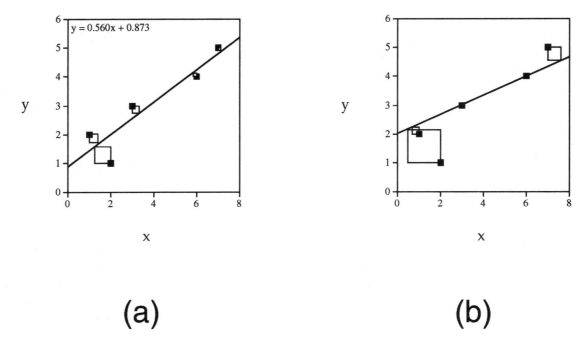

(a) **(b)**

Figure 10-5   *Linear fits to sample data (Table 10-1); (a) Least Squares fit, (b) alternate line through the data (not the best fit).*

   As with all things that your computer can do, line fitting was done first by hand. The fitting routine that spreadsheets and plotting programs use is simple enough that you can do it by hand if there aren't too many points. The (x,y) data for Figure 10-5 is listed in Table 10-1 along with the additional information needed to compute the slope and y-intercept for the best fit line. The symbol Σ (sigma) represents **summation**, adding all values of a particular type. You should be familiar with summation from taking an **average**, or **mean**, of a group of numbers. If there are **N** numbers with values x_i, the mean is the sum of all the numbers divided by **N**:

$$\text{mean}(x) = \frac{\sum_{i=1}^{N} x_i}{N}$$

The subscript i is an index for each number in the group. The summation is performed by adding all numbers, from $i = 1$ to $i = $ **N**. This is indicated below and above the summation sign (Σ). The index below the sigma is the starting number and the index above the sigma is where the series ends.

   For **n** data points ((x,y) pairs) the Least Squares equations are:

$$\text{slope} = m = \frac{n\sum x_i y_i - \sum x_i \sum y_i}{n\sum x_i^2 - (\sum x_i)^2}$$

Table 10-1   Data for Least Squares Method

	x	y	x²	xy
	1	2	1	2
	2	1	4	2
	3	3	9	9
	6	4	36	24
	7	5	49	35
Σ	19	15	99	72

$$\text{intercept} = b = \frac{\sum x_i^2 \sum y_i - \sum x_i \sum x_i y_i}{n \sum x_i^2 - (\sum x_i)^2}$$

Note that the denominator for both equations is the same. The subscripts i indicate that the sums encompass all data values in the series. When you calculate xy values you use the same point (hence the same subscript i) and multiply x•y. Many scientific calculators have the ability to perform a Least Squares fit to (x,y) data. With the equations above, you no longer have an excuse not to do it even if your calculator can't. A practice example is included at the end of this chapter.

An additional aspect of graphing experimental data involves proper treatment of error in your numbers. When a measurement is made there is inherent uncertainty in the number, and this uncertainty is expressed graphically as an **error bar**. The bar shows the range of uncertainty in the measurements, y ± error, which you can often estimate with reasonable accuracy. While the error bar is actually an "error box," since there is error in both the x and y dimensions, the error in the dependent variable is usually much greater than the error in the independent variable. For this reason it is customary to plot error in the y-dimension only, unless the error in x is large. Figure 10-6 (a) shows the data from Table 10-1 plotted using error bars in the y-dimension (error was chosen as ± 0.5 for illustration, detailed error analysis is actually a complicated issue and will not be treated here). Note that this extra information suggests that the point at (2,1) may be erroneous, since the best fit straight line does not pass through its error range. This result should cause the experimenter to reconsider that data point. If possible the measurement should be duplicated. If this is not possible (for instance if it is 3:00 a.m.—the morning your lab report is due), you can use the error bars to support elimination of a point from a large data pool. A new best fit line for the data, calculated without the erroneous data point, is shown in Figure 10-6 (b). Note that the slope and intercept for this new line is significantly different from that for the original data. With only a limited number of data points, it is advisable to remake measurements that have large deviations from the best fit line. However, if there are a large number of points then throwing out one or two has little effect on the overall quality of the data.

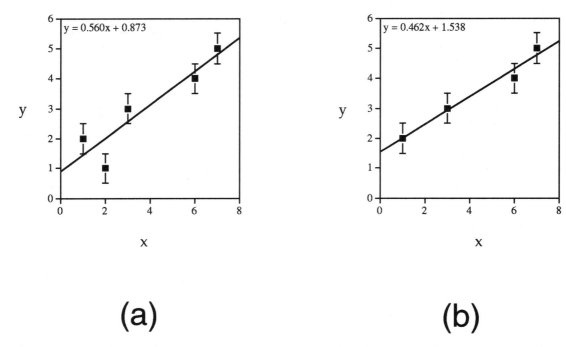

(a)   **(b)**

Figure 10-6 *Linear fits to sample data (Table 10-1) with arbitrary error bars; (a) Least Squares fit to all points, (b) improved Least Squares fit after elimination of "bad" point at (2,1).*

To create an effective graph you should pay attention to a few suggestions. It is likely that in the years to come you will be graphing more often with a computer, but sometimes there will arise the occasion to plot data the old fashioned way—on paper with a pencil. Most of these suggestions apply to both formats. Make the graph as large as possible on a sheet of paper, without crowding the axes against the margins. Use high quality graph paper with a fine rule (small divisions). Always label your axes clearly and with large letters. Include units on the axes. Choose divisions so that all of your data will fit on the page, and use spacing between divisions that is divisible by four or five. If the range of x or y (or both) is many orders of magnitude, change your scale from a linear scale to a logarithmic scale. Take the log of your numbers before plotting and label the axes appropriately. Plot points and include error bars if possible. Circle points to make them easier to see (tiny pencil points are hard to find on a graph). Make sure that your graph has a title that indicates the identity of the plot. Include derived information like slopes and intercepts. Always draw best fit lines with a high quality straight edge and a sharp pencil. When plotting functions (for example, a quadratic equation plots as a parabola) choose values of x that are close enough to give an accurate shape of the resultant curve. Near critical points like inflections, minima, or maxima, use values of x that are spaced closer together. If you follow these suggestions your results will be easy to read for both you and others that look over your data (like your instructor, your boss, etc.).

Chapter 10 Summary

- Common graphing methods include **bar charts**, **pie charts**, and **scatter plots**.

- In plotting data, the horizontal axis (**x-axis**) is used for the *independent variable* and the vertical **y-axis** (positioned perpendicular to the x-axis) is used for the *dependent variable*. Points have **(x,y) coordinates.**

- The equation for a line has the form $y=mx+b$, where m is the **slope** of the line and b is the **y-intercept**. The y-intercept is the value of y when $x = 0$ (where the line crosses the y-axis).

- The slope is the "rise over run," $m=\dfrac{"rise"}{"run"}=\dfrac{y_2-y_1}{x_2-x_1}=\dfrac{\Delta y}{\Delta x}$.

- The equation of a line can be determined from *any* two points on the line. From the equation, points that lie between data points can be determined by **interpolation** and those that lie outside of the data range by **extrapolation**. Never stake your reputation on an extrapolation!

- Relations of interest can be converted from their original form to a linear format so that measurements can be plotted as (x,y) pairs. Taking a root or logarithm of an equation is a common method of conversion.

- The "best fit" linear relation to data can be accomplished with the Least Squares Method.

- Uncertainty in measurements is expressed graphically as an **error bar**.

Practice Exercises Chapter 10

1. Plot the following data using the suggestions at the end of Chapter 10.

   a.

X	Y
0	0
1	2
2	4
3	5
4	6
5	6.5
6	7
7	7.5
8	8
9	8
10	8
15	8

b.

X	Y
1	2.4
2	3.0
4	4.1
5	4.5
6	5.0
8	6.1
10	6.9

2. Plot the lines with these equations:

   a. $y=2x+1$
   b. $x+2y+2=0$

3. Determine the slope and y-intercept for the lines in 1(b) and 2(b). Write the equation for the line in 1(b). (You must estimate the best line for 1(b) "by eye").

4. Using the graphs from above, determine a value for y at the following values of x:

   a. $x=3.0$ for line in 1(b)
   b. $x=14.0$ for line in 1(b)
   c. $x=-4.0$ for line in 2(a)
   d. $x=0.23$ for line in 2(a)
   e. $x=-2.0$ for line in 2(b)
   f. $x=22$ for line in 2(b)

5. Determine the Least Squares best fit line to the data in 1(b). Compare this equation to the one you drew "by eye" in problem #3.

Trigonometry and Geometry

"The sum of the square roots of any two sides of an isosceles triangle is equal to the square root of the remaining side!"

—The Scarecrow in The Wizard of Oz

Apparently the excitement of getting a new brain was too much for the Scarecrow. He mixed up his trigonometry and made an incorrect statement about isosceles triangles (triangles with two equal sides). He was on the right track, though. I think he meant to say "The sum of the equal sides of a *right* isosceles triangle is equal to the square root of the remaining side." Oh well, he did pretty well for someone with a brain made of straw.

Trigonometry involves relationships between the angles and sides of triangles. It may not be immediately obvious, but triangles are related to circles. Learning this connection will help you remember and use trigonometric functions. That skill will be valuable for solving problems in science which involve trigonometry and geometry.

Basic Trigonometry

Consider a circle with its center at the coordinates (0,0) and a **radius** (distance from center to outer edge) of exactly **1**. Imagine that you make the circle by rotating an arrow that has a length of **1** and lies initially on the positive x-axis, so that the tip rests at the coordinates (1,0). If you rotate the arrow through one complete revolution the tip travels the entire **circumference** of the circle, to return back to the starting position. The circle with a radius of 1 is called the **unit circle** ("unit" means "one"), and is shown in Figure 11-1. The circumference of the unit circle is exactly 2π ($C = 2\pi R$, $R = 1$). There are two measures of the position of the arrow. The first is the distance along the circumference the arrow has traveled away from the starting position. The second is the **angle** that the arrow makes with the positive x-axis.

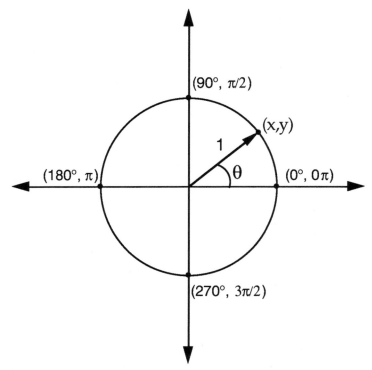

Figure 11-1 *The Unit Circle.*

The angle made between the arrow and the positive x-axis is often designated with the Greek letter θ (theta). This angle can be expressed in two ways, either using **degrees** or **radians**. 360° corresponds exactly to 2π radians. For the unit circle (only!) the angle measured in radians relative to the positive x-axis (θ = 0 radians and 0° for +x axis) is identical to the distance the radius arrow tip has moved along the circumference. All angles for all circles can be measured in either degrees or radians, with the conversion

$$\theta(\text{degrees}) \times \frac{2\pi \text{ radians}}{360°} = \theta(\text{radians})$$

For example, to convert θ = 90° to radians:

$$\theta = 90° \times \frac{2\pi \text{ radians}}{360°} = \frac{90°}{360°} \times 2\pi \text{ radians} = \left(\frac{1}{4}\right) \times 2\pi \text{ radians} = \frac{\pi}{2} \text{ radians}$$

It is a good idea to learn a few matches between the angle theta and the value of the radians (**n** represents a **whole number**, 0, 1, 2, ...):

0° = **0**, 2π, ... **n**(2π) radians

90° = **π/2**, π/2 +2π, ... π/2 +**n**(2π) radians

180° = **π**, π +2π, ... π +**n**(2π) radians

270° = **3π/2**, 3π/2 +2π, ... 3π/2 +**n**(2π) radians

360° = **0**, 2π, ... **n**(2π) radians

It should be easy to remember the values for 0°, 90°, 180°, 270°, and 360°. Intermediate values are pretty obvious, too, for instance 45° is π/4 radians.

Every point along the unit circle has coordinates, or (x,y) values. These values are related to the angle theta through the trigonometric functions **sine** and **cosine**. Consider the unit circle to have a radius R (equal to one in this case). The ratios of x and y to R depend on theta:

$$\frac{y}{R}=\sin\theta$$

$$\frac{x}{R}=\cos\theta$$

Since R = 1 for the unit circle, the x coordinate becomes cosθ, and the y coordinate becomes sinθ. The relation (x,y) = (cosθ,sinθ) and Figure 11-1 provide four points that are easy to remember: the points (1,0), (0,1), (-1,0), and (0,-1). You should be able to construct the entire circle using these points. I have only committed one to memory: (1,0) = (cos0°,sin0°). I can generate the rest of the unit circle knowing this "starting place."

The values of x and y can be plotted as a function of the angle. The result are periodic waves which undulate between values of -1 and +1. Figures 11-2 and 11-3 show plots of a sine wave and a cosine wave. Note the different values for where each wave crosses the x

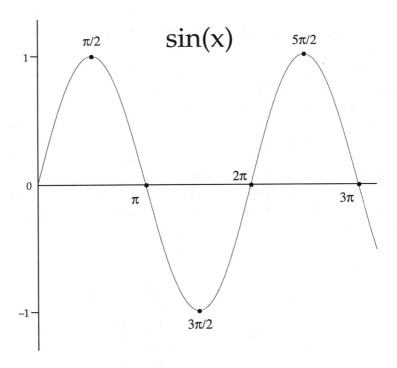

Figure 11-2 *Plot of y = sin(x).*

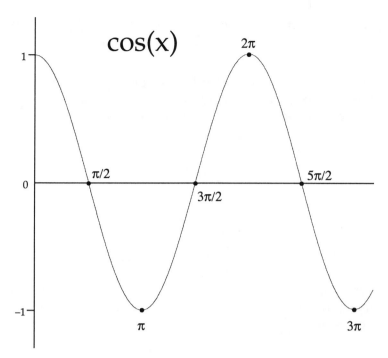

Figure 11-3 *Plot of y = cos(x).*

and y axes. These waves are nothing more than the values of (x,y) generated as you travel along the unit circle circumference. A circle with a radius not equal to one follows the same format. From the relations above, the coordinates along the circle become (x,y) = (Rcosθ, Rsinθ). The radius R gives the size of the circle, and when plotted as in Figures 11-2 and 11-3 it is the **amplitude** (height) of the sine wave and cosine wave.

How is this unit circle related to triangles? Note that the arrow shown in Figure 11-1 is the **hypotenuse** of a **right triangle** (contains one angle equal to 90°) formed with the positive x-axis. This is reproduced in Figure 11-4, alongside another right triangle with labels for the *hypotenuse* (H), the *opposite side* (O), and the *adjacent side* (A). These labels are chosen as they relate to the angle theta—O is the side opposite the angle theta, and A is the side adjacent to theta (but not the hypotenuse). A short phrase will help you remember the relations between the trigonometric functions sine, cosine, and **tangent** (the tangent is defined as $\tan\theta = \frac{\sin\theta}{\cos\theta}$): "**O**scar **H**ad **A** **H**eck **O**f **A** **T**ime **C**atching **S**almon." The first letters of each word represent **O**pposite, **H**ypotenuse, **A**djacent, **T**angent, **C**osine, and **S**ine, in the following equations:

$$\frac{\text{Oscar}}{\text{Had}} = \text{Salmon} \qquad \frac{\text{Opposite}}{\text{Hypotenuse}} = \sin\theta$$

$$\frac{\text{A}}{\text{Heck}} = \text{Catching} \qquad \frac{\text{Adjacent}}{\text{Hypotenuse}} = \cos\theta$$

$$\frac{\text{Of}}{\text{A}} = \text{Time} \qquad \frac{\text{Opposite}}{\text{Adjacent}} = \tan\theta$$

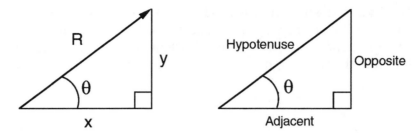

Figure 11-4 *Right triangles (see text for description of labels).*

Just read the sentence starting with Oscar and ending with Salmon (go down the left side and then up the right side). Stupid but effective.

An important relation between the values of the sides of a right triangle is the **Pythagorean Theorem**. If you know the value for two sides of a right triangle, you can find the length of the remaining side using this relation. Consistent with the labels used above, the Pythagorean Theorem becomes:

$$H^2 = O^2 + A^2$$

The square of the hypotenuse length is equal to the sum of the squares of the remaining sides. Sounds a lot like what the Scarecrow was trying to tell the Wizard, doesn't it? The Pythagorean Theorem is demonstrated in Figure 11-5, where the right triangle has legs

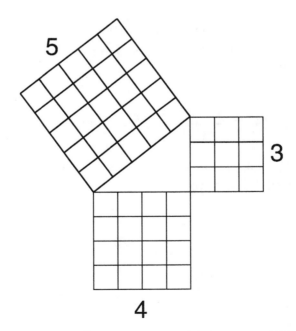

Figure 11-5 *Illustration of the Pythagorean Theorem. The square of the hypotenuse (25 blocks) is equal to the sum of the squares of the other sides (16 plus 9 blocks).*

with length 3 and 4 and the hypotenuse has length 5. You can see from the figure that $(3)^2 + (4)^2 = 9 + 16 = 25 = (5)^2$, the square of the hypotenuse length. An important consequence of the Pythagorean Theorem is a trigonometric identity that relates sine to cosine. If you apply this relation to the right triangle in our unit circle (Figure 11-4) then you find:

$$R^2 = y^2 + x^2$$

$$R^2 = (R\sin\theta)^2 + (R\cos\theta)^2$$

$$R^2 = R^2 (\sin^2\theta + \cos^2\theta)$$

$$1 = \sin^2\theta + \cos^2\theta$$

Note that in the identity, $\sin^2\theta = (\sin\theta)^2$ and $\cos^2\theta = (\cos\theta)^2$. This identity allows a quick calculation of $\sin\theta$ if you know $\cos\theta$, and vice versa.

Other trigonometric functions include **secant, cosecant,** and **cotangent**. These are the reciprocals of the functions sine, cosine and tangent:

$$\sec\theta = \frac{1}{\cos\theta}$$

$$\csc\theta = \frac{1}{\sin\theta}$$

$$\cot\theta = \frac{1}{\tan\theta}$$

The reciprocal trigonometric functions can be determined by taking the reciprocal $\left(\frac{1}{x}\right)$ of a function, for example

$$\sec(45°) = \frac{1}{\cos(45°)} = \sqrt{2} = 1.4142 \quad [\textbf{45 cos } \tfrac{1}{x}]$$

Note that $\sec\theta$ is *not* equal to $\cos^{-1}\theta$. Whereas $\sec\theta$ is a reciprocal function, $\cos^{-1}\theta$ is the **inverse** function of $\cos\theta$. For example,

$$\cos(45°) = \frac{1}{\sqrt{2}} = 0.7071$$

$$\cos^{-1}(\cos(45°)) = \cos^{-1}(0.7071)$$

$$45° = \cos^{-1}(0.7071)$$

The cos⁻¹θ will "undo" the action of cosθ, i.e., $\cos^{-1}(\cos\theta)=\theta$. The same holds for sin⁻¹θ and tan⁻¹θ as well:

$$\sin^{-1}(\sin\theta)=\theta \ \text{ and } \ \tan^{-1}(\tan\theta)=\theta$$

Consider the right triangle in Figure 11-4 with a hypotenuse of 20 ft and opposite side of 5 ft. Use the inverse sine function to find θ:

$$\sin\theta=\frac{5\text{ ft}}{20\text{ ft}}=\frac{1}{4}$$

$$\theta=\sin^{-1}\left(\frac{1}{4}\right)=14.5° \qquad [4 \ \frac{1}{x} \ \sin^{-1}]$$

When using trigonometric functions it is important to be aware of how your calculator treats numbers. Angles can be expressed in either radians or degrees, and it is critical that you set your calculator to operate in the mode that matches your input. This is usually changed using a "mode" key or key sequence. **Check your manual** to make sure you know how to operate your calculator in the proper mode. Some calculations will be easier to perform in degrees, while some formulas will require radians. Learn the difference and practice using your calculator in both modes.

Vector Basics

Look again at the arrow shown in Figure 11-1. It has a specific length (1 in this case) and **direction**. The direction is indicated by the angle theta relative to the positive x-axis. If you imagine that the unit circle is a compass with 0° corresponding to the direction of east, the arrow points generally toward the northeast (north is at 90°). These traits of length and direction are characteristics of vectors. A **vector** has a numerical value (with units) and a direction, such as 55 mph due west. Examples of vectors include electric and magnetic fields, forces like gravity, dipole moment, acceleration, and velocity. A **scalar** has only a numerical value and unit, like $5, 3 inches, or 250 pounds.

One way to indicate a vector is to place an arrow over a variable, such as $\overline{A}$. The vector $\overline{A}$ has a **magnitude** $|A|$, where the bars indicate to take the absolute (positive) value of the vector length. Figure 11-6a shows a vector $\overline{A}$ as it relates to x and y axes. By dropping lines from the arrow tip perpendicular to the axes you perform "projections" of the vector onto the axes. The lengths of the projections are related to the vector by the Pythagorean Theorem (refer to Figure 11-6):

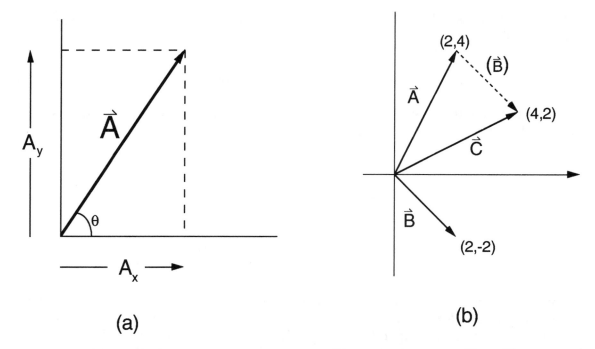

Figure 11-6 *Characteristics of vectors; (a) projections of vector $\overline{A}$, (b) addition of vectors $\overline{A}$ and $\overline{B}$.*

$$A_x = |A|\cos\theta$$
$$A_y = |A|\sin\theta$$
$$|A| = \sqrt{A_x^2 + A_y^2}$$
$$\theta = \tan^{-1}\left(\frac{A_y}{A_x}\right)$$

Scalars and vectors can be multiplied together to change the length and possibly the direction of the vector. For the product $s \times \overline{A}$, if the scalar s is positive then the resultant vector has length $s|A|$ and points in the *same* direction as $\overline{A}$. If s is negative then the resultant vector has length $s|A|$ but points in the *opposite* direction ($-\overline{A}$). For example,

$$-2 \times (30 \text{ mph east}) = -(60 \text{ mph east}) = 60 \text{ mph west}$$

The order does not matter for addition of vectors, they are said to **commute**:

$$\overline{A} + \overline{B} = \overline{B} + \overline{A}$$

It also does not matter how additions are grouped, vectors are **associative**:

$$(\overline{A}+\overline{B})+\overline{C}=\overline{A}+(\overline{B}+\overline{C})$$

To subtract vector $\overline{B}$ from $\overline{A}$, reverse the direction of $\overline{B}$ and add to $\overline{A}$:

$$\overline{A}-\overline{B}=\overline{A}+(-\overline{B})$$

Notice that when vectors of equal magnitude and opposite direction are added the resultant is equal to zero:

$$\overline{A}+(-\overline{A})=\overline{A}-\overline{A}=0$$

When adding vectors, the tip of one vector is placed at the tail of the other (see Figure 11-6b). The vector $\overline{A}$ can be represented by the coordinates of the tip (2,4). Vector $\overline{B}$ has tip coordinates (2,-2), and its tail is placed at the tip of vector $\overline{A}$. The sum $\overline{A} + \overline{B}$ results in $\overline{C}$, found by adding the x values and the y values:

$$\overline{C}=\overline{A}+\overline{B}=(2,4)+(2,-2)=(2+2,4-2)=(4,2)$$

The other way to look at this is with the projections:

$$\overline{A} : A_x =2, A_y =4$$
$$\overline{B} : B_x =2, B_y =-2$$
$$\overline{C} : A_x +B_x , A_y +B_y =2+2,4-2=(4,2)$$

Therefore,

$$C_x =4, C_y =2 \text{ and } |C|=\sqrt{4^2 +2^2} =\sqrt{20}$$

Basic Geometry

Most of the geometry that you will use in first courses of science will involve simple structures like rectangles, circles, and triangles. It is important that you learn the relations that govern the boundaries of these structures. You are expected to know the basic characteristics of regular shapes, including **perimeter**, **area**, and **volume**.

The **perimeter** (has units of length) is the distance around the outside of a closed shape. For straight-sided shapes you must add the lengths of the sides. A rectangle has a length (L) and a width (W). The perimeter is simply 2L + 2W. For a triangle you add the lengths of the three sides to determine the perimeter. A circle has only one side—the perimeter is called the **circumference**. The circumference of a circle is equal to the **diameter** (distance across the circle, measured through the center) times pi. Since the diameter is twice the radius, this relation becomes $C = 2\pi R$. Refer to Figure 11-7 for a summary of simple geometry of rectangles, circles, and triangles.

The **area** (has units of length squared) of a regular shape is how much space the shape occupies in two dimensions. The area of a rectangle is just the length times the width. The area of a circle is πR^2 (my ninth grade math teacher said that if you can remember that "pie are round" you shouldn't forget that "πR^2." Hint: say it out loud). The area of a triangle is a little tricky. For a right triangle (one 90° angle) the area is simply one half of the product of the sides that are not the hypotenuse (O and A from above). You can see this if you imagine putting two right triangles together so that their hypotenuses touch. The result is a rectangle with length and width determined by the two legs of the original triangle. Since the area of the rectangle is $L \times W$, and the triangle is exactly half the area of the rectangle (you made the rectangle by adding two triangles), the area of the triangle must be $\frac{1}{2}L \times W$. For the triangle we call L and W the Height and the Base. The area of a triangle is $\frac{1}{2}$(Base×Height). Refer to Figure 11-7 for an example that is not a right triangle. Note that all that is required is to pick one side as the base and draw a vertical line to the apex of the triangle. The length of this vertical line is the Height of the triangle. This vertical line creates two right triangles "back to back." The area is determined as above, only this time you make two smaller rectangles from the newly drawn right triangles.

The previous discussion applied to two-dimensional regular shapes. In three dimensions, shapes occupy **volume** (has units of length cubed). Consider a rectangle that you transform into a box. The box has length, width, and depth. The volume is simply the combination of these factors, $V = L \times W \times D$. (Sometimes depth is expressed as height, $V = L \times W \times H$.) You can make two three dimensional structures from a circle—a sphere and a cylinder. A sphere has a volume of:

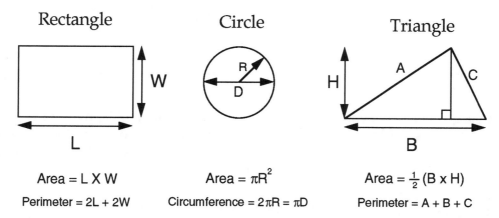

Figure 11-7 *Summary of geometric characteristics for rectangles, circles, and triangles.*

$$V_{sphere} = \tfrac{4}{3}\pi R^3$$

A cylinder is a circle "extruded" in one direction so that it has a height, H. The volume of the cylinder is this height multiplied by the area of the circular base:

$$V_{cylinder} = H \times \pi R^2$$

If a triangle is extruded in one direction the result is a prism, again with a new dimension, D. Like the cylinder, the prism volume is the area of the triangle multiplied by the depth of the prism:

$$V_{prism} = \tfrac{1}{2}(Base \times Height) \times D$$

Depth was chosen here instead of height to reduce confusion with the height of the triangle used to calculate the area. Note in all cases that the units of volume involve a length cubed (like cm³ or in³), and that area has units of length squared (cm², in², etc.).

Chapter 11 Summary

- Trigonometric functions are related to circles and triangles.

- A circle has a radius and contains 360° (2π radians). The unit circle is used to generate the trigonometric functions **sine** and **cosine**. The ratios of x and y to the radius depend on theta through the functions sine and cosine. The angle 0° corresponds to the point (1,0) = (cos0°,sin0°) on the unit circle.

- For a circle with a radius other than one, coordinates along the circle become (x,y) = (Rcosθ, Rsinθ), where R is the **amplitude** of the sine and cosine functions.

- A **right triangle** has a hypotenuse opposite the right angle (90°).

- The **Pythagorean Theorem** relates the square of the hypotenuse of a right triangle to the squares of the remaining two sides.

- From the Pythagorean Theorem, the identity $1 = \sin^2\theta + \cos^2\theta$ can be derived.

- Be careful to set your calculator to operate in the mode that matches whether you calculate in degrees or radians.

- **Vectors** have **magnitude** and **direction**. A vector $\overline{A}$ that makes angle θ with the x-axis has magnitude

$$|A| = \sqrt{A_x^2 + A_y^2},$$

where $A_x = |A|\cos\theta$ and $A_y = |A|\sin\theta$. The angle θ is found from

$$\theta = \tan^{-1}\left(\frac{A_y}{A_x}\right).$$

Scalars have magnitude but no direction.

- To add vectors, sum the x and y projections: $(C_x, C_y) = (A_x + B_x, A_y + B_y)$ for $\overline{C} = \overline{A} + \overline{B}$.
- Characteristics of regular shapes which you should learn, include **perimeters, areas,** and **volumes**.

Practice Exercises Chapter 11

1. Convert these angles in **radians** to angles in **degrees**:

   a. π c. $-3\pi/2$
   b. $\pi/4$ d. 2.43

2. Convert these angles in **degrees** to angles in **radians**.

   a. $135°$ c. $30°$
   b. $-270°$ d. $257.3°$

3. Compute these trigonometric functions (only use your calculator when necessary):

   a. $\sin(2\pi)$ c. $\sin(45°)$
   b. $\cos(-\pi/2)$ d. $\tan(60°)$

4. Use your skill with trig functions to determine the following for a right triangle:

   a. For a 23° angle, the adjacent side = 2.5. How long is the hypotenuse?

   b. For an unknown angle, the adjacent side = 3.0 and the hypotenuse is 4.0. What is the angle?

   c. For an unknown angle, the adjacent side = 2.2 and the opposite side = 4.4. How long is the hypotenuse?

   d. The triangle has a 62° angle and a hypotenuse = 5.0. What are the lengths of the other two legs of the triangle? What are the values of the other two angles?

5. Find the length of the third side of these right triangles:

   a. Hypotenuse = 5, Leg (1) = 4

   b. Hypotenuse = 4, Leg (1) = 3

   c. Leg (1) = 2, Leg (2) = 1

   d. Leg (1) = 3, Hypotenuse = 6.

6. Given the following information, determine the requested quantity:

   a. $\sin(\theta) = 0.8290$, $\theta = ?$

   b. $\cos(\theta) = -0.6018$, $\cos(\theta) = ?$

   c. $\tan(\theta) = 1.00$, $\sin(\theta) = ?$

   d. $\cos^2(45°) = ?$

7. a. What angle is made between the x-axis and a vector that originates at (0,0) and terminates at (4,3)?

   b. Add the vectors $\overline{A}$ and $\overline{B}$ if $\overline{A} = (-2,2)$ and $\overline{B} = (4,1)$.

   c. Determine $\overline{A} - \overline{B}$ for the vectors in 7(b).

8. Determine the perimeter, area, or volume where appropriate:

   a. Calculate the perimeter and area for a square with a side = 4.2 cm.

   b. Calculate the area of a circle with a diameter of 6.0 in.

   c. Calculate the volume of a box with dimensions $2 \times 4 \times 5$.

   d. Calculate the volume of a cylinder with radius = 2.0 and height = 10.0

   e. Calculate the volume of a spherical droplet of water with radius of 1.0 micrometer.

   f. Calculate the area of a triangle with base length 10.0 in and height 2.00 in.

Solving Problems

Entire books have been written on the subject of how to solve problems in math and science courses. It would be impossible to summarize that large body of knowledge in this one chapter. Instead, I will focus on some basic strategies that might help you approach problems so that you have a good chance of success. You become a good problem solver only after much practice and hard work (sorry to burst your bubble). Solving problems quickly and correctly depends on your knowledge of the subject, your skill with mathematical techniques, and the insights that you gain by doing many and different problems. Anyone can be a good problem solver! Start now to build your skills.

When faced with a problem to solve, the single best piece of advice is: look before you leap. It doesn't hurt to know what you are doing, either. What that means is, if you learn to proceed slowly with the confidence that comes from knowing your subject, you have a much better chance of determining the right answer. You must **read** the problem carefully (don't skim it or make any assumptions as you go). A common complaint of students after an exam is that they just didn't read the problems carefully enough to know what to do to get them right. Often students try to solve the problem without really understanding what information is given and what they need to determine as the answer. Use this basic approach for any problem:

1. Read the problem word for word.

2. As you read, write down information given in the problem. Some prefer to underline or circle words and symbols in the problem, but I find it useful to write down variables and what they are equal to. Give quantities specific names, like v(initial) and v(final) for initial and final velocities, respectively. Draw a little diagram if it helps.

3. Write down explicitly what information is being sought. Most often this is one variable that will have a particular value. Write down the anticipated units for the answer.

4. Write down the relationship between the information you know and the desired answer. This will usually be an equation in a familiar form.

5. Rearrange the equation to a convenient form to allow you to solve for the desired parameter. Put the numbers you know into the formula. Doing this early will identify conversions that you need.

6. Always carry units in the calculation, and make appropriate unit conversions so that your answer has the proper units.

7. Check your answer to make sure it is reasonable (more on making estimates in Chapter 13). Write down your answer using correct units and significant figures.

Here is an example to try with the above approach in mind:

> How many feet does a swallow fly in twenty minutes at an average air speed of 15 miles per hour?

Write down what you know: $t = 20$ min $v = 15$ mph

(It is tempting to write down "swallow." The type of bird has no real bearing on the problem so don't write it down, it will only waste time.)

Write down what you seek: distance, $d = ?$ ft

Write down a relation that you know: speed = distance/time

$$v = d/t$$

Rearrange to solve for distance: $d = v \cdot t$

Insert numbers: $d = (15\frac{\text{mi}}{\text{hr}})(20\,\text{min})$

Change units with conversions: $d = (15\frac{\text{mi}}{\text{hr}})(5280\frac{\text{ft}}{\text{mi}})(20\text{min})(\frac{1\,\text{hr}}{60\,\text{min}})$

Perform calculation: $d = 5\,\text{mi}\,(5280\frac{\text{ft}}{\text{mi}}) = 53{,}000\,\text{ft}$

Check your answer: Five miles is a long way, but it seems reasonable. A million miles wouldn't be reasonable.

Note that in the example above, information was always attached to a variable and the units were written down explicitly. The problem asked for the distance in feet, so it was necessary to convert the units of speed (velocity, v) from miles per hour to feet per hour. If the units had not been carried, then that important change might not have been made and the problem would have been answered incorrectly. Carry the units to avoid simple mistakes like that.

A subtle, yet very important, aspect of problem solving is being familiar with your subject enough to think about it in terms of the variables commonly used in calculations. If you try to cram before an exam then you are less likely to get mathematical problems correct because you have not familiarized your mind with the equations and their variables. You must expose yourself to variables multiple times before you begin to think about scientific relations in terms of variables and equations. This is why it is better to study in smaller amounts every day than to try to learn most of the material right before an exam. Through constant exposure, you train yourself to recognize and value the information that appears in problems that you will solve. By doing exercises on your own (Ack! Homework!) you help develop the familiarity that you need to solve problems on exams.

Most problems require the use of one equation to solve them. With one equation you can determine only one unknown quantity. For every unknown quantity you must have an equation—two equations for two unknowns for instance. Advanced problems might require you to determine one quantity from one relation and then input that result into another equation to get the answer to the problem. Multistep problems will come easier to you after you familiarize yourself with the individual steps. The strategy of writing down what you know and what you seek will help you see early whether there are intermediate steps in the problem which must be solved. Purposefully expose yourself to these types of problems. They build your abilities by strengthening connection of ideas through the use of equations.

Sometimes it will seem like you don't have the information that you need to solve a problem. This can happen when there are constants in the equations that you use to solve problems. The constants themselves are not given in the problem, so you have to look them up or remember them. It is a good idea to ask your instructor what constants you are assumed to know for an exam (commonly they are given in a table at the beginning or end of an exam). Another instance, when it appears that you need more information, is when you have neglected a simple relation like the one between percentages: all percentages add to a total of 100% (also, all fractions must add to a total of 1). If you forget this simple relation you might get stuck while doing a problem.

When you hit a brick wall like this, you must have the experience behind you to guide your actions. A problem on an exam is never impossible, you just might be "challenged." Maybe you have never put two equations together before, or you haven't done a percentage problem where one of the percentages was missing. The experience of homework (there's that word again) will help you out of those jams and allow you to "think on your feet" during an exam. Build those strengths now by doing many problems and watching the problem solving techniques of others—your instructor, teaching assistants, tutors, or class-mates. Look for tricks and shortcuts only *after* you have convinced yourself that you can do it the "long way" first. Once you really understand what you are doing you will see the shortcuts while practicing problems on your own (homework, homework, homework!).

Chapter 12 Summary

- Read problems carefully and write down what you know, what you seek, and what relation (equation) will get you there.

- Always carry units (sometimes the units of the desired result will suggest the equation that you need to use).

- Check your answer to make sure it is reasonable. If you have time, put the numbers through it twice to make sure you get the same answer.

- Familiarize yourself with variables and equations to build your ability to do problems (do your homework!).

- Every unknown needs an equation, and there will only be one unknown per equation—you must have a value for every other variable and constant.

- Practice, practice, practice.

Practice Exercises Chapter 12

1. A student finds a gold-colored, heavy rock having a mass of 22.0 g. The student fills a graduated cylinder with water to exactly 10.0 cm³. When the rock is placed in the water, the new volume in the cylinder is 14.3 cm³. The density of pure gold varies between 15.3 and 19.3 g/cm^3. Is this rock a piece of gold?

2. Laser light pulses from certain Nd:YAG lasers last only about 5 ns. If the speed of light is 2.998 x 10⁸ m·s⁻¹, what is the length of a single laser pulse? If the pulse is shortened to 2 ps, how long is it?

3. Marbles with radii of 5.0 mm are glued side-by-side around the circumference of an earth mover tire. The tire has a diameter of 10.0 ft. How many marbles encircle the tire?

4. The speed of sound in air is 1140 ft/s (at 25°C.). If lightning strikes exactly 5.0 miles from you, how long will it take to hear the thunder? Determine a general equation to help you determine the distance of lightning strikes from a measure of the time between lightning flash and thunder clap.

5. In 1924 the French physicist Louis de Broglie made the somewhat startling suggestion that *all* moving particles, from electrons to city buses, have a wavelength, λ. That wavelength is determined from the constant h (6.63 x 10⁻³⁴ J·s), the particle mass m, and the velocity v:

$$\lambda = \frac{h}{mv}$$

What is the wavelength of a 150 g baseball thrown by a pitcher at 90 mph? (Hint: remember that 1 J = 1 kg·m²·s⁻²).

6. You want to wallpaper the 9 ft × 9 ft wall in your bedroom. The rolls of wallpaper are 18 in wide and 20 ft long. How many rolls do you need?

7. An ice skating rink is 200 ft × 50 ft, covered with a 1.0 in layer of ice. What is the total weight of ice in pounds if the density of the ice is 0.92 g/cm^3?

8. Lemon Biscuits:

300 g flour	1 egg
180 g butter	1 pinch salt
70 g sugar	1 grated lemon rind
1 tsp vanilla	Bake 15-20 min. at 170-190°C.

Convert this recipe to weights in ounces, assuming there are 28 g in one ounce. Convert the oven temperature to Fahrenheit.

9. Air pressure at sea level is approximately 14.7 psi (pounds per square inch). What is the maximum weight of argon gas in a column of air with a base of one square mile? The composition of air is 75.5% nitrogen and 23.1% oxygen.

10. In the surveyor's chain system of measurements, 1 link = 7.92 in, 1 chain = 100 links, and 1 furlong = 10 chains. The Kentucky Derby is a horse race of 10 furlongs. How many miles is this?

CHAPTER 13

Making Estimates

After you have mastered a basic understanding of how to solve problems, it is advantageous to learn how to make estimates. Estimates are useful in a number of situations. You should be able to estimate your answers to make sure that the number determined with your calculator is reasonable. You can usually spot calculator errors when the calculated result is orders of magnitude different than your estimate. Estimates can be useful on multiple choice tests where selections often differ by more than 10%. A quick estimate can be faster than punching a calculator, and just as good if all you are trying to do is match a selection listed below the problem.

The key to good estimating is rounding numbers up or down to give values that are easier to multiply or divide. You have probably done this many times without thinking about applying it to scientific problems. An example might be planning for a party. Suppose you expect about twenty people over for chips, dips, and soft drinks, and you want to estimate how much money the party will cost. Assume that everyone will drink two drinks, or 40 total, so seven six packs will cover them. Five big bags of chips and three containers of dip will be enough (as long as they have already had dinner!). You saw drinks on sale for $1.89 a six pack, so you estimate seven at $2 for a total of $14. Five bags of chips at $2 per bag (really $1.99) is $10, and the dip is about $3 per container (really $2.89) for another $9. The estimated sum is $14 + $10 + $9 = $35. The real cost is $31.85, so you overestimated, but only by about ten percent—enough to cover the sales tax. Clearly the estimate took less time than the real calculation and gave a number that was close enough to be useful.

You can do the same thing with calculations used in solving problems. The most important thing to practice is quick estimates using numbers expressed in scientific notation. If you can't multiply or divide exponential numbers without your calculator, then you won't be able to make good estimates. Estimate an answer to the following:

How far away is the sun if it takes sunlight 8.3 minutes to reach earth?

What you know: $t = 8.3$ min $v = c$ (speed of light) $\approx 3 \times 10^8$ m/s

What you seek: distance, d = ? meters

Relation: $v = d/t$

Rearrange: $d = v \cdot t$

Estimate:

$$d = (3 \times 10^8 \tfrac{m}{s})(8\text{min})(60 \tfrac{s}{\text{min}}) = (24 \times 60)(10^8)\, m = (144 \times 10)(10^8) = 1440 \times 10^8 \, m$$

This answer is equal to 1440×10^5 km, or 144×10^6 km (144 million kilometers). If this had been a multiple choice test with the following selections,

A) 10 million km   D) 150 million km

B) 50 million km   E) 200 million km

C) 100 million km

which would you choose as the right answer? The estimate took less time than real calculator crunching, and yielded the right answer (D).

Here is another example with more gymnastic manipulations, an equilibrium calculation with numbers in exponential notation. For the formation of hydrogen iodide, an equilibrium expression is determined to be:

$$K_C = \frac{[HI]^2}{[H_2][I_2]} = \frac{(1.6482 \times 10^{-2})^2}{(2.9070 \times 10^{-3})(1.7069 \times 10^{-3})}$$

The first order of business is to separate the exponents so that the power of the result can be determined:

$$K_C \cong \frac{(1.65)^2}{(2.91)(1.71)} \bullet \frac{(10^{-2})^2}{(10^{-3})(10^{-3})}$$

Note that the numbers were reduced to three significant figures. Now the exponents can be reduced, and we should look for obvious factors in the numerator and denominator. Since 1.65 is close to 1.71, we will eliminate 1.65 once from the numerator and remove 1.71 from the denominator:

$$K_C \cong \frac{1.65}{2.91} \bullet \frac{10^{-4}}{10^{-6}} \cong \frac{1.65}{3} \bullet (10^2)$$

The denominator can be rounded to three for the division into 1.65 . This is quick to do long hand, but you can estimate it by comparing some nearby "perfect" divisions. For example, 1.5/3 is 0.5 and 1.8/3 is 0.6, so the value of 1.65/3 will be half way between these results, namely 0.55 . Our estimate becomes:

$$K_C \cong 0.55(10^2)=55$$

This was wildly successful, since the answer determined with five significant figures on a calculator is 54.748! Practice with these methods will cut down on steps and the time it takes to make manipulations.

When dealing with fractions, you want to reduce their complexity by dividing the numerator and denominator by common factors. Speed is related to how well you know your "times tables." You need to see common factors quickly and make simplifications. If you depend on your calculator to do simple manipulations like 3×14, or $28 \div 7$, then you need to practice doing more calculations "in your head."

Estimating roots and logarithms takes some additional practice. Roots more complex than square roots are difficult, but square roots are not too bad. For example, what is the square root of 70? Seventy is between the perfect squares 64 (8^2) and 81 (9^2), so the square root is between 8 and 9. Since 70 is less than halfway between 64 and 81, you might guess that the square root is less than 8.5, say about 8.3. That is a good guess since $\sqrt{70}$ = 8.37 (no, I didn't cheat with my calculator, I really estimated 8.3).

You can use a similar approach with logarithms (base ten logs; natural logs aren't so easy). What is log(779)? In scientific notation 779 is written 7.79×10^2. Since 779 is between 100 (or 10^2) and 1000 (or 10^3), you know that log(779) is between 2, the log of 100, and 3, the log of 1000. The log scale is not linear, so interpolating between the extremes (2 and 3 in this example) is tricky. You will want to learn one more piece of information: the **log of 5.0 is 0.70** to two significant figures (log(2.0) = 0.30 and log(8.0) = 0.90 are also handy). Therefore:

$$log(500) = log(5.0 \times 10^2) = log(5.0) + log(10^2) = 0.70 + 2 = 2.70$$

Similarly,

$$log(0.000050) = log(5.0 \times 10^{-5}) = log(5.0) + log(10^{-5}) = 0.70 + (-5) = -4.30$$

Since 779 is between 500 and 1000, we estimate that log(779) is between 2.7 and 3. A good guess might be 2.85 . That is a little low, since the correct answer is 2.89, but it was an excellent estimate (again, no calculator! Honest!).

Making estimates that are useful takes practice. If you have never tried making estimates then it is time that you begin. The best time to try is when you are doing homework problems. Put your calculator aside and try to do the problems first by making estimates, then check your answer with the calculator. Follow along with examples in your text but make estimates in the margin to match the numbers used by the textbook author. After a few weeks of this you will be a champion estimator, and you will strengthen your abilities to predict answers to problems without picking up your calculator. This is a valuable skill! Some of the "insights" that your instructors want you to gain in their courses involve making quick judgments and predictions. This type of thinking comes from understanding the material and knowing that the numbers "look right." Wean yourself off of your calculator! Good estimators are good students. Make both of these your goals.

Chapter 13 Summary

- Learn how to estimate your answers to make sure that the number determined with your calculator is reasonable and to develop a "feel" for the numbers.

- Make good rounding decisions to give values that are easier to multiply or divide in a complicated calculation.

- Good estimators multiply or divide exponential numbers quickly without a calculator.

- Practice with these methods will cut down on steps and the time it takes to make estimates.

- Brush up on your "times tables" if you have forgotten them.

- Interpolating (guessing values between obvious extremes) in estimating roots and logarithms take practice.

- The best time to practice is when you are doing homework problems. Start now and you will be rewarded.

Practice Exercises Chapter 13

1. Estimate the weight in tons (1 ton = 2000 lbs) of bauxite in a train car with approximate dimensions of 5 ft × 5 ft × 10 ft. Bauxite is an aluminum ore with a density of about 2.5 g/cm^3.

2. Estimate the energy of a 158 grain bullet traveling with a muzzle velocity of 2500 ^{ft}/s. Use the kinetic energy equation:

$$E_{kinetic} = \tfrac{1}{2}\, mv^2$$

Express your answer in kJ (1J = 1 kg·m^2·s^{-2}). One pound is equal to 7000 grain.

3. Vinegar is a solution of acetic acid with a hydrogen ion concentration, [H$^+$], near 0.004 mol/L. Estimate the pH of vinegar, knowing the definition

$$pH = -\log [H^+]$$

4. A 1 ft^3 plexiglas box is filled with chili beans. You measure 20 beans visible in a square inch of the box. Estimate the number of beans in the box to win a free trip to the East Texas Chili Festival.

5. Estimate your gasoline costs for a road trip to New Orleans for Mardi Gras, starting in Detroit, Michigan (a 1078 mile trip one-way). Your car gets about 28 mpg. Gas in your area sells for about $1.08 per gallon. If you maintain an average speed of 65 mph, and only make four twenty-minute pit stops, what is your total travel time (down and back)?

6. How may laps (down and back) in a 25-meter pool equals one mile of swimming?

CHAPTER 14

Some Study Tips

How do you study? Does it work? Could you do different things to help you learn better? Your answers are probably something like: "1. Well, I really don't know. 2. Yeah, sure, most of the time. 3. I guess so, but what?" I don't pretend to have the answers for you. I do have some observations and some suggestions that might help.

One of the biggest problems facing students and educators is that students have many different learning styles. What's worse, your instructors most probably have a *different* learning style than you do. That means that your instructor won't be making a special effort to present material in class that is customized to the way in which you might learn it best. The bottom line is this: your performance in a class depends more on your personal relationship with the material than on the behavior of your instructor. It may be true that a good instructor might inspire you, challenge you, or really interest you in a subject. These are positive attributes that can help you maintain your motivation in class, but they will not take the place of good studying and hard work on your part.

So back to studying. What is it? It is your interaction with the material to be learned in a class. If you have little interaction, you will most likely learn only a little. If you have a lot of interaction you *are not guaranteed* to learn a lot, however. Often students "study" many hours before an exam, yet do poorly because the way they studied did not promote understanding of the material. Commonly this means that they exposed themselves to individual tasks or skills, but never added enough depth to their studying to allow them to actually learn the material. They knew the parts but could never see the whole.

In this chapter I will list for you some things related to studying that might be a help. Some of them will seem like common sense, but are included here because you can never have too much common sense. The others will be ideas that have worked in the past for others, which you may adopt as you see fit. Remember, there is no "secret" or "magic bullet." It is likely that you will spend years developing study habits that work for you.

Maybe some of the ideas presented here will help accelerate that process. Read on and try anything that looks like it might help.

Here comes the common sense part:

1. **Read your textbook**. Really read it, every word. I often have students complain about exam questions, then look very sheepish when I show them the exact problem from the text. Get in the habit of reading for **understanding**, not just exposure. Students who "skim" or "look over" the text are usually D students or worse. Take notes while you read—with a pencil in a separate notebook. Throw away the highlighter! Marking up your book with obnoxious day-glo inks will not help you learn better. More on this later.

2. **Go to class.** Not only will you see the material again, it will be through someone else's eyes. And that someone is very important. Your instructor interprets the course material and naturally stresses the most important points in class. Use this time to gather additional information about where you should place emphasis in your study of the material. Often favorite examples or problems worked in class will reappear on exams. Listen closely to your instructors because they are attempting to help you make connections about ideas in the course—developing your depth of understanding. Don't go to sleep! If you find yourself slipping away, get more sleep the night before or concentrate harder on your notes. Try to predict what will be said next, or work on practice problems—do anything to stay awake.

3. **Take notes.** It will be very difficult to remember the wonderful exciting things said by your instructors, unless you write them down. Listen for voice inflections and write down phrases when your instructors seem to emphasize things. Copy closely when derivations are done or equations are manipulated. If your instructor writes something on the board you should write, too. But don't focus so much on writing that you miss what is said. Develop a shorthand (don't attempt to take down things word for word) so you can write the important information while *listening* to presentation of ideas. If this is hard for you, try taping the lectures (ask permission first) and writing only things that are written on the board.

4. **Do your homework.** Sometimes it is not clear what is expected of you with regard to homework. Homework might not be collected for grading. Usually problems are selected by your instructor but they might not be covered in class. The fact of the matter is that you are expected to do homework, since it represents an active way for you to learn the material. Just because you never turn it in or your instructor rarely talks about it, don't make a mistake by assuming that it is not important. It is very important. You might find (no mystery here) that your instructor will often ask exam questions that are very similar to exercises in the text which appear as examples or homework.

5. **Stay ahead of your instructor.** Read your text before you attend class. That way the words used by your instructor won't be new, and the lectures will help you connect ideas that you have read about. When you read you might write down things that are most difficult for you. Then when you attend your lecture you have prepared your

mind for the explanations. If the explanations do not materialize then refer to suggestion 6.

6. **See your instructor.** If you read about an idea, hear that idea again in class, look it over after class, and if it still doesn't make sense to you—go see your instructor. Instructors set aside time to meet with their students (office hours). Often a few minutes with your instructor can clear up little misunderstandings before they blossom into deep problems with the material. For some reason students ignore this option and don't show up at their instructor's office until it is usually too late to reverse any damage done. Your instructor wants you to succeed, to see you understand the material. Take your list (see #5) to your instructor for help. Your instructor can help you best when there are specific problems to address and work through.

7. **It makes a difference where you study.** If you are distracted while you study then you will have a harder time retaining material. Don't watch television. Use quiet music (quiet, not necessarily boring) in the background if you need noise to keep you company. If you are constantly interrupted by friends dropping by or calling on the phone then you will have to leave. Find a place to spread out and get comfortable: a library, student lounge, or fast food restaurant (during off hours when traffic is light). Above all make sure that the place you pick to study helps you concentrate on reading and working problems. It can be a chore to find such a place, but you will be rewarded by deeper understanding.

8. **It makes a difference how much time you spend studying.** This suggestion is not exactly what it seems. Certainly if you study only a few hours right before an exam you will not be rewarded with a good grade. No, this recommendation has more to do with the length and timing of your study sessions. For example, make sure that you study your hardest subject *first*. It is only natural to put off to the end of the night your work on the most difficult material. Unfortunately, by the time you get to your difficult studying you are the most tired and least likely to learn efficiently.

The most important thing is to study in small bursts. Study each subject *every day*. Read and work through assignments in short sessions (30 minutes to an hour and a half each). A big mistake made often is to "save up" time for a subject and then try to cram too much learning into one or two marathon sessions before an exam. If you read a little every day and keep ahead of your instructor, you will find it much easier to comprehend the subject and "keep up" in class. Never delude yourself that you can catch up later when you have no clue about what your instructor is talking about in class. That feeling is a danger sign that should jump-start your studying efforts.

It is wise to write down how you spend *every* hour of the week, and then schedule time to study particular subjects. You will be surprised how many hours you waste between classes or events. Use these odd bits of time to your advantage by carrying notes with you to look over and "distill" (more on distilling below). How much study time is reasonable? As a rule of thumb, for every hour you spend in class per week you should spend two hours studying (at a minimum). If you have a four credit science course then this means eight hours a week outside of class (more if the course is hard and you are struggling).

9. **Studying with others can help.** By talking over problems with other students in your class you will use the new words and ideas that you are learning. This will help strengthen your understanding of the subject. It is important, however, that you study properly with others. Too often group study situations turn into opportunities for socializing. You must do things to avoid this. One step is to pick a neutral site for studying that is not conducive to socializing (a cafeteria table is a good place). Plan to go somewhere together after your study time to satisfy your urge to socialize and reward yourselves for sticking to the job of studying. Be careful not to rush through the material in order to do the socializing, however.

It is best if you can structure your activities by deciding on specific tasks to accomplish between meetings. Preparing short practice exams (to be discussed more fully below) is one aid that could work for your group. Another possibility is to pick the most important topics from the text and assign members of your study group to summarize them during the meetings. These short summaries can initiate discussion of aspects that others feel a need to talk about. The person that took the time to prepare the summary will most likely be in the best position to help others with their problems. In this way every member of the group becomes a resource. The group becomes a meeting of tutors!

10. **A tutor can help.** But why fork out the money until you absolutely have to? Often students turn to help from tutors before they have spent good study time on their own. If you have tried the ideas listed above, but still need some personal training, then getting a tutor should be a possibility. But don't get a tutor until help during your instructor's scheduled office hours is not enough. Your instructor should be most willing to help you and you should not pay for help until the two of you have determined that it is a good option. Remember that if you communicate with your instructor then the two of you can work together to make your study efforts pay off. Your instructor might feel comfortable suggesting someone to contact for tutoring. It is usually a violation of school policy for your instructor to offer to tutor you for money. If this happens, talk to the head of the department. A good use of a tutor is to hire someone for your study group (you can split the cost!). Do this right before an exam to help clear up last minute problems with the material.

11. **Avoid some "freshman" mistakes.** The first year of college can be a trying time. It might be your first time away from home. You are now responsible for your own time table. There is no one but you to decide when things must be done. Scheduling time for adequate study can be a challenge, especially since you are surrounded by new and interesting things: new people to meet, new places to explore, new ideas and activities. It is a good idea to get involved with your surroundings, but the wise student needs to strike a balance. You might not know how much extracurricular activity you can handle without jeopardizing your grades. The best approach is to take it slowly. Don't try every new thing that appeals to you. Treat your work like a 40 hour a week job. Make sure that the sum of your time in class and your study time is at least 40 hours per week. If your school work is getting done easily then you can add other activities into your schedule. Avoid new things that require a great deal of time. Be critical of your need for a part time job. Only work if you must to pay for your

education. Working thirty hours a week to pay your car insurance is a burden, and a threat to good grades (this is a real example from a real student).

Once you establish your ability to do your class work, socialize, and engage in other activities, you will be more able to judge how much time each new activity will take. In the beginning, however, just stick to your class work and a small amount of socializing until you get your study habits established. Establishing study habits can be hard, especially for someone who did not study much during his/her high school years. It is important to realize that college level material goes faster and there is less structured help for you in class. You are on your own to rise to the challenge, and if you can see that before you get started then you might avoid the biggest "freshman" mistake: wishful thinking. Students with poor study backgrounds often stumble on their first exam and then write it off to bad luck. They predict that things will get better. Of course they rarely do, unless a serious effort is made to bring study habits up to speed. A bad performance on an exam should sound an alarm to you. See your instructor and restructure your studying to get you back on track. The situation will never take care of itself, it is up to *you* to make things happen.

12. **Do anything that helps**. Take advantage of all the resources available to you. Investigate solution manuals and software, particularly if recommended by your instructor or a previous successful student of the course. Go to help sessions and encourage your classmates to go as well. Larger attendance at help sessions will generate more and better questions for your instructor. I can't emphasize too much the advice to go see your instructor for help. In a private meeting your instructor can take more time to explain things that might have gone by you too fast in class.

Note Distillation

There are two other specific activities that may help you study better. The first involves a new way of reading. The second involves studying with at least one other person. Both have yielded positive results when employed diligently by students. In fact, dramatic improvement has taken place for students who have changed their old habits in part by adopting these new approaches. They might work for you, too.

Remember the admonishment to throw away your highlighting pen? Highlighting material in your text does not cause your brain to absorb it any better. What is worse, most students highlight *too much*, so that when you go back over your text you are faced with pages of glowing yellow paragraphs. It just doesn't work. But if you are more selective you will learn more. Instead of reading with a highlighter, keep a separate notebook from the one that contains your class notes. In this notebook write things down as you read. Don't write complete sentences, use phrases instead. Write brief definitions that paraphrase the ones in the text (use your *own* words). Write down formulas and define variables clearly but not elaborately. This will only take you a little longer than reading without taking notes, and it will cause the material to be "seen" more by your brain. You have to process the information more deeply if you are going to write something down.

If a chapter in your text is about twenty pages long, you might have handwritten notes of ten pages. Your goal is to "distill" those pages down to one page. In the end you want one page per chapter to study right before your exam. The best time to distill your notes is during those odd times between classes—during lunch, while you are waiting to talk with your instructor, or whenever you have a few minutes. Always carry the notebook with you since you never know when spare moments will arise.

The purpose of distilling your notes is simple: when you take those ten handwritten pages and distill them to five, you do so by only writing down the things that give you the most difficulty. In other words, you will not write down the things you **know**—the things you have learned. By doing this a few times, you will convert that ten pages into one page that contains only the things you have the most trouble remembering. It might take you four iterations, but they will be surprisingly short. And you will generate a customized study guide. By exam time you may find that you know your study sheets by heart, which is a wonderful situation.

This method really works. My most dramatic example is a student who made a 58 on the first exam in General Chemistry and then made a 95 on the second exam. This exciting turnaround came about by using the note distillation method, coupled with a few additional study hours (mostly between classes). That student learned a different way of interacting with the material and could absorb it better using this reading technique. This new approach probably helped her in her future courses as well.

Mock Exams

Another study method that might work for you requires that you find another student willing to work with you (helpful hint: don't choose a boyfriend or girlfriend). It can work for a study group, too. Your task is to write exam questions. Arrange with your partner or group to meet in a neutral, boring place once a week, or whatever is appropriate for the coverage in your class. Agree to write a certain number of questions over the material covered in the textbook. A good number might be ten questions per chapter. If your class exams will be multiple choice then make your questions multiple choice. Try to match the way you will be examined for your course.

At the appointed time, exchange practice exams with your partner(s). Give yourselves a reasonable amount of time to do the questions, but no more than you will be allowed for the real exam. After time is called, grade the practice exams. For the problems that are missed, the author must explain the right answers for the benefit of the others in the group. After all the problems have been covered, adjourn to the local café for a bite to eat (if you reward yourself for studying, your Pavlovian instincts will help keep you motivated).

What is the purpose of writing exam questions? Quite simply, if you can write a good question that measures whether someone knows the material then you must know that material pretty well yourself. This is just a sly way to get you to learn the material well enough so that you can write a legitimate question about it. You also have the added benefit of taking a practice exam to show how well you understand the material. Try to write good questions over the most important aspects of the material. It might be hard at first, and you will likely get frustrated. Try using the practice examples and homework problems from your text as a guide when you first get started. With practice you will be able to write questions more easily. Remember, the idea behind this method is to get you and your partners more involved in active learning of the material. Writing questions can definitely help.

Weird Tricks

Finally, let me pass along some weird tricks that I know about, but don't necessarily know why they work—or if they work very often. **Weird Trick #1:** (This tip sounds absolutely stupid, but some students claim success, and I can't argue with success.) Listen to music the night before an exam (but not the same thing over and over!). During the exam you can "play back" the songs in your head and some of the material might come back to you easier (I told you it was weird). Studies show that students who study in a certain environment will perform better on tests given under similar circumstances. For example, students who study surrounded by the smell of chocolate will do slightly better on an exam given in a room which has the same smell. This implies that if you study in the room in which your exam will be given you might have a slight edge during exam time. It's worth a try. **Weird Trick #2:** Dress up for the exam. Some students report that if they dress well for an exam, the extra self confidence that comes from looking good helps them relax and perform better. This isn't like having a lucky pair of socks, it's more like having a lucky evening gown. Give it a try. I have students that swear by this one. **Weird Trick #3:** Read a novel the night before your final exam. By that time cramming won't help anyway, so the best thing you can do is relax. Get a good night's sleep. Don't worry with the material the night before the big exam, but look over your study sheets right before you walk into the exam room.

Most of all, change things that don't work. Stay on top of your studies, and when you feel uncomfortable, get help! The way you are studying might not be effective, and you need to modify activities accordingly. Don't try to do it all on your own if you can't seem to do well enough without help. Your instructors really do want you to succeed (otherwise they shouldn't be teaching). Seek help from them as soon as you know that things are not progressing properly.

Chapter 14 Summary

- Your performance in a class depends more on *you* than on your instructor.
- Read your textbook and take notes while you read (but throw away your highlighter!).
- Go to class and take notes.

- Do your homework and write down things that are most difficult for you to help your instructors help you when you go to their office hours.

- Study in small bursts, hardest subject first.

- Use group study effectively (write practice exams or assign summaries to group members).

- Enlist a tutor only after you have exhausted your own arsenal of study aids. Follow the advice of your instructor in this regard.

- Avoid wishful thinking—seek help when you feel unsure of your progress.

- Try distilling your notes to one page per chapter to generate customized study sheets.

- Write exam questions to learn the material in more depth. Trade your questions with a partner to help each other.

"Should I Buy a Computer?"

In the first week of classes each fall semester a few students ask me the following question: "Should I buy a computer?" My answer is usually "It depends." A short discussion follows, which boils down to the following: Computer technology changes so fast that whatever you buy will not be state-of-the-art in six months, so don't shoot for the slickest, fastest, most expensive computer (yet). In the first year of college the heaviest use for computers by students is for word processing and simple graphing that can be accomplished with spreadsheet programs. The really exciting stuff—sound and video, for instance—won't be necessary for success in your courses. It will be later in your college career that you will use advanced programs that require real speed, but at that time you will be granted access to computers that will allow you to work on these programs. So what you really need is a low end machine with a printer so you can bang out those English 101 essays. Wait, you could do the same with nothing but a typewriter!

Well, not really. Face it, today the typewriter has been replaced by the word processor. With word processing you can edit your work easily, tinker with it until it really sings. And typewriters can't perform spreadsheet calculations. It is probably worth the extra money to gain that freedom. The ability to create and access information with computers make them so powerful. But what type of computer? What platform (operating system) should you choose?

This is a sticky question, particularly since the issue of platform is changing so rapidly. My best advice is to find out what most people on your campus use. Go to the Computer Center and ask around. Find out if there is a campus computer network that you can access by either a port (permanent connection like a phone line) in your dorm room or a dial-in connection using a modem. Get a machine that is compatible with the campus network, because there is usually a stock of free programs available through this network. Ask about using the **Internet** (a national computer network that allows global access) and **electronic mail**. Make sure that your computer will allow you to tie in to the campus mainframe computer system (and hence the electronic mail system). When you are e-mail active, send me a short note about this book (see the Preface for my e-mail address). Above all, see if you can get a special deal on a computer through the Computer Center. Often the university allows you to take advantage of educational discounts. You can save big bucks by waiting until you arrive on campus to purchase your machine.

Don't head into a computer purchase too fast. Evaluate all the options first. Remember, it is perfectly acceptable to hand in a handwritten English essay! In addition, it is likely that your college has computers available in computer labs. You can take a few minutes to word process your latest masterpiece on equipment purchased by the school. It will only cost you the price of a floppy disk. Be wary of advice from your friends that use computers. People typically choose one brand and stick with it. If you are new to computers it is probably best for you to "shop around" and use different computers before you buy. You will most likely have access to different computers on campus. Seek them out.

The bottom line on a computer is: wait as long as possible (you save money and get a better machine this way) and don't waste money on speed and glitz that you don't need. Remember—you aren't buying an arcade game, you're buying a tool to help save you time and improve your learning. So don't overspend on bells and whistles that you won't need for your studies. A simple machine with a good word processor and spreadsheet will do. Splurge on a color monitor, though, since a monochrome monitor might put you to sleep while working on your lab report at 2:00 a.m.

Answers to Practice Exercises

Chapter 1 Answers to Practice Exercises

Here are possible approaches. There are other key sequences that will give the right answer.

1. $\left[\,3.53-6.2=\mathbf{Mem}\right.$ put denominator into memory

   $3.53+6.2=\div\mathbf{Rec}=\left.\right]$ recall denominator from memory

   Answer: -3.64

2. $\left[\,4\times3.5=-6=\mathbf{1/_x}\,\right]$ use inverse key on denominator

   Answer: 0.125

3. $\left[\,6.2-2.6=\log\times2.33\,\mathbf{Mem}\right.$ put second term into memory

   $4.12\,\mathbf{x}^{\mathbf{y}}\ 3=-\mathbf{Rec}=\left.\right]$ use x^y key to do the cube

   Answer: 68.6

4. $\left[\,4\times6=\mathbf{Mem}\,5\,\mathbf{x}^2-\mathbf{Rec}=\sqrt{}+5=\right]$

   Answer: 6

5. $\left[\,6.5\,\mathbf{EXP}7\pm\mathbf{Mem}\right.$ put denominator into memory

   $4.7\,\mathbf{EXP}3\pm+2.3\,\mathbf{EXP}3\pm=\div\mathbf{Rec}=\left.\right]$

   Answer: $10769 = 1.1\times10^4$ (See chapters 6 and 7 for discussion of scientific notation and significant digits.)

6. $\left[3\times5=\text{Mem}\,3x^2-\text{Rec}+6=\right]$

  Answer: 0

7. $\left[0.1-5\,\text{EXP}\,2\pm=\text{Mem}\right.$ put first part of denominator into memory

   $0.1+5\,\text{EXP}\,2\pm=\times\text{Rec}=\text{Mem}$ put denominator into memory

   $2\times5\,\text{EXP}\,2\pm=x^2$ square the numerator

   $\div\text{Rec}=\left.\right]$

  Answer: 1.3

8. $\left[8.31\times298=\right.$ calculate denominator of exponent first

   $\div5000\,{1}\!/\!{x}$ divide denominator by numerator and then **take reciprocal**

   $\pm\,e^x\left.\right]$ change sign on exponent before computing e^x

  Answer: 0.133

9. mode select **radians**, then

   $\left[\pi\div2=\text{Mem}\right.$ put $\left(\pi/2\right)$ into memory since it is used twice

   $\sin+\text{Rec}\cos=\left.\right]$ take $\sin\left(\pi/2\right)$ immediately after putting $\left(\pi/2\right)$ into memory

  Answer: 1

10. mode select **degrees**, then

   $\left[1.00\tan^{-1}\right]$ or $\left[1.00\,\text{INV}\,\tan\right]$

  Answer: 45°

Chapter 2 Answers to Practice Exercises

1. a. $5{5}/{9}=\dfrac{(9)(5)+5}{9}=\dfrac{45+5}{9}=\dfrac{50}{9}=5.5\bar{5}$

The bar over the 5 in this answer is a **repetend**, which indicates that the 5 repeats indefinitely (your calculator will display a number of digits and round off the last one).

$$[5 \div 9 = +5 =] \text{ yields } 5.5\overline{5}$$

b. $2^{23}/_{32} = \dfrac{(32)(2)+23}{32} = \dfrac{64+23}{32} = \dfrac{87}{32} = 2.71875$

$$[23 \div 32 = +2 =]$$

c. $3^{5}/_{11} = \dfrac{(11)(3)+5}{11} = \dfrac{33+5}{11} = \dfrac{38}{11} = 3.45\overline{45}$

$$[5 \div 11 = +3 =]$$

d. $-4^{2}/_{7} = -\left(\dfrac{(7)(4)+2}{7}\right) = -\left(\dfrac{28+2}{7}\right) = -\dfrac{30}{7} = -4.28571$

$$[2 \div 7 = +4 = \pm \ ]$$

2. a. $\dfrac{13}{8} = \dfrac{8+5}{8} = 1 + \dfrac{5}{8} = 1^{5}/_{8} = 1.625$

$$[13 \div 8 =]$$

b. $\dfrac{305}{137} = \dfrac{2(137)+31}{137} = 2^{31}/_{137} = 2.22628$

$$[305 \div 137 =]$$

c. $-\dfrac{56}{26} = -\dfrac{28}{13} = -\left(\dfrac{2(13)+2}{13}\right) = -2^{2}/_{13} = -2.15385$

$$[56 \div 26 = \pm]$$

d. $\dfrac{256}{16} = \dfrac{(16)(16)}{16} = 16$

3.

a. $^{1}/_{8} = 0.125$ b. $^{1}/_{3} = 0.3\overline{3}$

$^{3}/_{8} = 0.375$ $^{2}/_{3} = 0.6\overline{6}$

$^{5}/_{8} = 0.625$ $^{4}/_{3} = 1.3\overline{3}$ $(1^{1}/_{3})$

$^{7}/_{8} = 0.875$ $^{5}/_{3} = 1.6\overline{6}$ $(1^{2}/_{3})$

4. a. $13.375 = 13 + 0.375 = 13 + \dfrac{3}{8} = 13\tfrac{3}{8} = {}^{107}\!/_8$

b. $7.625 = 7 + 0.625 = 7 + \dfrac{5}{8} = 7\tfrac{5}{8} = \dfrac{61}{8}$

c. $-4.3\overline{3} = -\left(4 + 0.3\overline{3}\right) = -\left(4 + \dfrac{1}{3}\right) = -{}^{13}\!/_3$

d. $3.6\overline{6} = 3 + 0.6\overline{6} = 3 + {}^{2}\!/_3 = 3\tfrac{2}{3} = {}^{11}\!/_3$

5. π: Your calculator should have a key that returns π when pressed (it may be a secondary function that requires an extra key to be pressed). If not, you can get π as the inverse cosine of -1,

$$\left[1 \pm \cos^{-1}\right] \text{ or } \left[1 \pm \text{INV}\cos\right]$$

Make sure your calculator mode is set to **radians**.

$$\pi \cong 3.141592654$$

e: Determine **e** as e^1

$$\left[1e^x\right] \qquad e \cong 2.718281828$$

Chapter 3 Answers to Practice Exercises

1. a. $20:120 = 2:12 = 1:6$

$$\frac{20}{20+120} = \frac{20}{140} = \frac{1}{7}$$

b. $13:52 = 1:4$

$$\frac{13}{13+52} = \frac{13}{65} = \frac{1}{5}$$

c. $2.4:17.2 = 1.2:8.6 = 0.6:4.3 = 1:7.1\overline{6}$

$$\frac{2.4}{2.4+17.2} = \frac{2.4}{19.6} = 0.1224$$

d. $2:12{,}000 = 1:6{,}000$

$$\frac{2}{2+12{,}000} = \frac{2.4}{12{,}002} = 0.0001666$$

Note that $\dfrac{2}{12{,}000} = 0.000166\overline{6}$, so $2:12{,}000$ is very nearly equal to $\dfrac{2}{12{,}000}$ since $12{,}000$ is so much larger than 2.

2. Home Fans: Visiting Fans$=45,000:5,000=45:5=9:1$

$$\%\,\text{Visitors}=\frac{\text{visitors}}{\text{visitors}+\text{home fans}}=\frac{5,000}{5,000+45,000}$$

$$=\frac{5,000}{50,000}=\frac{1}{10}\times100\%=0.10\times100\%=10\%$$

Note that a ratio of 1:9 is 10%, since 1:9 converts to $\dfrac{1}{1+9}=\dfrac{1}{10}$

3. Reptiles = 22%

Amphibians = 100% - 22% = 78% are amphibians

Number of amphibians = (78%)(74 animals)

= 0.78(74) = 57.72 = 58

The number is rounded up since there cannot be fractions of animals.

58 amphibians$=^{58}\!/_{74}=0.78\ =78\%$

$\underline{16\ \text{reptiles}}\quad =^{16}\!/_{74}=0.22\ =\underline{22\%}$

74 animals $\qquad\qquad\qquad$ 100%

4. $\dfrac{4}{5}\big(632\ \text{dentists}\big)=505.6=506\ \text{dentists}$

$\dfrac{4}{5}=0.8=80\%$

Minority opinion dentists = 100% - 80% = 20%

5. $\big(50\ \text{weeks}\big)\big(40\ \text{hours per week}\big)=\big(50\ \text{weeks}\big)\left(40\dfrac{\text{hours}}{\text{week}}\right)=2000\ \text{hours}$

$$\frac{\$37,500}{1\,\text{year}}=\frac{\$37,500}{50\,\text{weeks}}=\frac{\$750}{1\,\text{week}}=\$750\,\text{per week}$$

$$\frac{\$37,500}{1\,\text{year}}=\frac{\$37,500}{2000\,\text{hours}}=\frac{\$18.75}{1\,\text{hour}}=\$18.75\,\text{per hour}$$

6. $\%\,\text{impurity}=100\%-99.44\%=0.56\%=\dfrac{0.56}{100}$

$$\frac{0.56}{100}(2000\,\text{lb})=11.2\,\text{lb impurities in one ton soap}$$

7. $\frac{3}{8}$ inch $=0.375$ inch

$$\text{Elongation}=\frac{0.375\,\text{inch}}{120\,\text{inches}}=0.003125$$

$0.003125\times100\%=0.3125\%$

$0.003125\times1000\,\text{ppt}=3.125\,\text{ppt}\,(\text{a better format})$

8. $20\,\text{ppm}=\dfrac{20}{1,000,000}$

$$20\,\text{ppm}(10,000,000\,\text{gal})=\frac{20}{1,000,000}(10,000,000\,\text{gal})$$
$$=\frac{20\times10\times1,000,000\,\text{gal}}{1,000,000}=20\times10\,\text{gal}=200\,\text{gal}$$

200 gallons of PCB's are dissolved in the reservoir.

Chapter 4 Answers to Practice Exercises

1. a. $100,000=10^5$

b. $512=2^9$

c. $0.0001=10^{-4}$

d. $625=5^4$

2. a. $7^4=7\cdot7\cdot7\cdot7=49\cdot49=2401$ $\left[7x^y\ 4=\right]$

b. $(1.5)^3=(1.5)(1.5)(1.5)=(2.25)(1.5)=3.375$ $\left[1.5x^y\ 3=\right]$

c. $10^7=10\cdot10\cdot10\cdot10\cdot10\cdot10\cdot10=10,000,000$ $\left[10x^y\ 7=\right]$

d. $2^{-2}=\dfrac{1}{(2\cdot2)}=\tfrac{1}{4}=0.25$ $\left[2x^y\ 2\pm=\right]$

e. $\sqrt[3]{125}=\sqrt[3]{5^3}=5$ $\left[125x^{1/y}\ 3=\right]$

f. $(121)^{1/2}=\sqrt{121}=\sqrt{(11)^2}=11$ $\left[121\sqrt{\ }\right]$

3. a. 51.53632 $\qquad$ $\left[2.2\mathbf{x}^y\ 5=\right]$

b. 0.0356 $\qquad$ $\left[5.3\mathbf{x}^y\ 2\pm=\right]$

c. 0.5 $\qquad$ $\left[8\mathbf{x}^{1/}y\ 3\pm=\right]$

d. 1.162 $\qquad$ $\left[\pi\div2=\mathbf{x}^{1/}y\ 3\ =\right]$

e. 0.718 $\qquad$ $\left[4.21\mathbf{x}^y\ 0.23\pm=\right]$

f. 0.0432 $\qquad$ $\left[\pi\pm\mathbf{e}^x\right]$

4. a. $2^3+2^4=8+16=24$

b. $10^3-10^2=1000-100=900$

c. $\sqrt{\pi^3+\pi}=5.844$ $\qquad$ $\left[\pi\mathbf{x}^y\ 3=+\pi=\sqrt{}\right]$

d. $\dfrac{10-10^2}{10+10^2}=\dfrac{10-100}{10+100}=\dfrac{-90}{110}=-0.81\overline{81}$

5. a. $\left(10^3\right)\left(10^{-2}\right)=10^{3+(-2)}=10^{3-2}=10^1=10$

b. $\dfrac{10^{-2}}{10^4}=10^{-2-4}=10^{-6}$ or $\dfrac{10^{-2}}{10^4}=10^{-2}\left(10^{-4}\right)=10^{-2-4}=10^{-6}$

c. $\left(2^{-\pi}\right)\left(2^{-1}\right)=\left(\dfrac{1}{2^{\pi}}\right)\left(\dfrac{1}{2}\right)=\dfrac{1}{2\left(2^{\pi}\right)}=\dfrac{1}{2^{\pi+1}}=2^{-(\pi+1)}$

$\left(2^{-\pi}\right)\left(2^{-1}\right)=2^{-\pi+(-1)}=2^{-\pi-1}=2^{-(\pi+1)}=0.0567$

$\left[\pi+1=\pm\mathbf{Mem}\,2\mathbf{x}^y\ \mathbf{Rec}=\right]$

d. $\left(10^5\right)\left(10^{-1/5}\right)=10^{5-1/5}=10^{4.8}=63096$

$\left[5\,{}^{1/}\!\mathbf{x}\pm+5=10^x\right]$

It is simpler to recognize that $\frac{1}{5}$ is equal to 0.2, so that $5-\frac{1}{5}=5-0.2=4.8$. The keystroke sequence simplifies to

$\left[4.8\,10^x\right]$

Chapter 5 Answers to Practice Exercises ————————

1. a. $\log(1000) = \log(10^3) = 3$

b. $\log(0.00001) = \log(10^{-5}) = -5$

c. $\log(8) = 0.903 \quad [\mathbf{8 log}]$

d. $\log(0.23) = -0.638 \quad [\mathbf{0.23 log}]$

e. $\ln(e^3) = 3$

f. $\ln(2) = 0.693 \quad [\mathbf{2 ln}]$

2. a. $\text{antilog}(-4) = 10^{-4} = 0.0001$

b. $10^{-2} = 0.01$

c. $10^{\log(3)} = 3$

d. $e^{\ln(2)} = 2$

e. $e^{-1} = \frac{1}{e} = 0.368 \qquad [\mathbf{1 \pm e^x}]$

f. $e^{2\ln(3)} = e^{\ln(3^2)} = e^{\ln(9)} = 9$

3. a. $\log(2) + \log(4) = \log(2 \cdot 4) = \log(8) = 0.903$

b. $\log(3^2) = 2\log(3) = 0.954$

c. $\ln(10^{-2}) = -2\ln(10) = -4.605$

d. $\log(4) - \log(2) = \log(\frac{4}{2}) = \log(2) = 0.301$

e. $\log(10 \cdot 100) = \log(10) + \log(100) = 1 + 2 = 3$

f. $\log(4 + 3) = \log(7) = 0.845$

4. $\ln(x) = \ln(10) \cdot \log(x) = 2.303 \log(x)$

a. $\log(2) = \dfrac{\ln(2)}{\ln(10)} = \dfrac{0.693}{2.303} = 0.301$

b. $\log(e) = \dfrac{\ln(e)}{\ln(10)} = \dfrac{1}{\ln(10)} = \dfrac{1}{2.303} = 0.434$

c. $\log(100) = \dfrac{\ln(100)}{\ln(10)} = \dfrac{\ln(10^2)}{\ln(10)} = \dfrac{2\ln(10)}{\ln(10)} = 2$

Chapter 6 Answers to Practice Exercises

These answers are presented without regard to significant figures (See Chapter 7).

1. a. $54{,}000.00 = 5.4 \times 10^4$

b. $142.35 = 1.4235 \times 10^2$

c. $-0.00131 = -1.31 \times 10^{-3}$

d. $0.00000004 = 4 \times 10^{-8}$

2. a. $1.6 \times 10^{-4} = 0.00016$

b. $0.2 \times 10^3 = 200$

c. $3.7542 \times 10^3 = 3754.2$

d. $4.0 \times 10^6 = 4{,}000{,}000$

3. a. $\left(3.4 \times 10^{-4}\right)\left(1.7 \times 10^3\right) = (3.4)(1.7) \times 10^{-4+3} = 5.78 \times 10^{-1} = 0.578$

$$\left[\,3.4\,\mathbf{EXP}\,4\pm\times1.7\,\mathbf{EXP}\,3=\,\right]$$

b. $\dfrac{3.4 \times 10^{-4}}{1.7 \times 10^3} = \left(\dfrac{3.4}{1.7}\right) \times 10^{-4-3} = 2.0 \times 10^{-7}$

$$\left[\,3.4\,\mathbf{EXP}\,4\pm\div1.7\,\mathbf{EXP}\,3=\,\right]$$

c. $\left(0.2 \times 10^{-4}\right)\left(0.4 \times 10^{-4}\right) = (0.2)(0.4) \times 10^{-4-4} = 0.08 \times 10^{-8} = 8 \times 10^{-10}$

$$\left[\,0.2\,\mathbf{EXP}\,4\pm\times0.4\,\mathbf{EXP}\,4\pm=\,\right]$$

d. $\dfrac{5.0 \times 10^6}{-2.5 \times 10^{-5}} = \left(\dfrac{5.0}{-2.5}\right) \times 10^{6-(-5)} = -2.0 \times 10^{11}$

$$\left[\,5.0\,\mathbf{EXP}\,6\div2.5\pm\mathbf{EXP}\,5\pm=\,\right]$$

4. a. $\left(2.2 \times 10^{-2}\right)+\left(1.4 \times 10^{-2}\right) = (2.2+1.4) \times 10^{-2} = 3.6 \times 10^{-2}$

$$\left[\,2.2\,\mathbf{EXP}\,2\pm+1.4\,\mathbf{EXP}\,2\pm=\,\right]$$

b. $\left(5.6\times10^4\right)-\left(2.1\times10^4\right)=\left(5.6-2.1\right)\times10^4=3.5\times10^4$

$\left[5.6\,\text{EXP}\,4-2.1\,\text{EXP}\,4=\right]$

c. $\left(4.53\times10^2\right)-\left(0.2\times10^3\right)=\left(4.53\times10^2\right)-\left(2\times10^2\right)=\left(4.53-2\right)\times10^2=2.53\times10^2$

$\left[4.53\,\text{EXP}\,2-0.2\,\text{EXP}\,3=\right]$

d. $\left(1.1\times10^{-2}\right)-\left(3.0\times10^2\right)=0.011-300=-299.989$

$\left[1.1\,\text{EXP}\,2\pm-3.0\,\text{EXP}\,2=\right]$

Chapter 7 Answers to Practice Exercises

1. a. 5 sig figs

b. 4 sig figs

c. 4 sig figs

d. 7 sig figs

2. a. $2.74450=2.74$

b. $532.62=533=5.33\times10^2$

c. $0.13251=0.133$

d. $4223.0=4220=4.22\times10^3$

3. a. $\left(3.33\right)\left(6.4290\right)=21.4$

b. $\left(7.00\right)\left(8.3\right)=5.8\times10^1$

c. $\dfrac{4.0\times10^2}{0.2}=2000=0.2\times10^4$

d. $\left(-0.05009\right)\left(1.2\times10^8\right)=-6.0\times10^6$

e. $78.834-78.8=0.0\left(\text{one digit after the decimal!}\right)$

f. $\dfrac{0.03303}{454.34}=7.270\times10^{-5}$

g. $\ln\left(321.2\right)=5.7721$

h. $2.70+\log\left(0.0020\right)=2.70+\left(-2.70\right)=0.00$

Chapter 8 Answers to Practice Exercises

Some intermediate steps are shown.

1. a. $42=14x$

$$x=\frac{42}{14}=\frac{21}{7}=3$$

b. $x+2=2(x-2)=2x-4$

$2+4=2x-x$

$6=x$

c. $3x^2=75$

$x^2=\frac{75}{3}=25$

$x=\sqrt{25}=5$

d. $3x=2(x+1)=2x+2$

$3x-2x=2$

$x=2$

2. a. $2x+2y=x^2$

$2y=x^2-2x$

$y=\frac{1}{2}x^2-x$

b. $E=RT\ln\left(\frac{1}{y}\right)$

$\frac{E}{RT}=\ln\left(\frac{1}{y}\right)=\ln\left(y^{-1}\right)=-\ln y$

$\ln y=-\frac{E}{RT}$

$y=e^{-\frac{E}{RT}}$

c. $2-y=3a-x$

$-y=3a-x-2$

$y=-(3a-x-2)=2+x-3a$

d. $2x^2 - x - 1 = 4xy - 2y^2$

$2x^2 - 4xy + 2y^2 = x + 1$

$x^2 - 2xy + y^2 = \dfrac{x+1}{2}$

$(x-y)^2 = \dfrac{x+1}{2}$

$x - y = \sqrt{\dfrac{x+1}{2}}$

$y = x - \sqrt{\dfrac{x+1}{2}}$

e. $2\ln(x) + \ln(y) = 2$

$\ln(x^2) + \ln(y) = 2$

$\ln(yx^2) = 2$

$yx^2 = e^2$

$y = \dfrac{e^2}{x^2} = \left(\dfrac{e}{x}\right)^2$

f. $y + 1 = 2(x+1) = 2x + 2$

$y = 2x + 2 - 1 = 2x + 1$

3. a. $y + 5 = 3(x+1)$

$y = 3x + 3 - 5 = 3x - 2$

b. $k = Ae^{-2x}$

$\ln(k) = \ln(A) + \ln\left(e^{-2x}\right) = \ln(A) - 2x$

$\ln(k) = -2x + \ln(A)$

if $y = \ln(k), m = -2, b = \ln(A)$

c. $\dfrac{E}{h} = v - v_o$

$E = hv - hv_o$

if $y = E$, $x = v$, then $m = h$ and $b = -hv_o$

d. $2y=\dfrac{x^2-2x+1}{x-1}=\dfrac{(x-1)^2}{x-1}=x-1$

   $y=\dfrac{x-1}{2}=\frac{1}{2}x-\frac{1}{2}$ $m=\frac{1}{2},b=-\frac{1}{2}$

4. a. $4x-x^2-4=0$

   $x^2-4x+4=0$ $x=\dfrac{4\pm\sqrt{16-4(1)(4)}}{2(1)}=\frac{4}{2}=2$

   $(x-2)^2=0$

   $x=2$

   b. $x=\frac{6}{x}-1$

   $x^2=6-x$

   $x^2+x-6=0$ $x=\dfrac{-1\pm\sqrt{1-4(1)(-6)}}{2(1)}=\dfrac{-1\pm\sqrt{25}}{2}=\dfrac{-1\pm5}{2}=\dfrac{4}{2},\dfrac{-6}{2}=2,-3$

   $(x+3)(x-2)=0$

   $x=-3,+2$

   c. $x^2+4x+3=0$ $x=\dfrac{-4\pm\sqrt{16-4(1)(3)}}{2(1)}=\dfrac{-4\pm\sqrt{4}}{2}=\dfrac{-4\pm2}{2}=\dfrac{-2}{2},\dfrac{-6}{2}=-1,-3$

   $(x+1)(x+3)=0$

   $x=-1,-3$

   d. $x(1+x)=0.75$

   $x-x^2=\frac{3}{4}$

   $4x+4x^2=3$

   $4x^2+4x-3=0$ $x=\dfrac{-4\pm\sqrt{16-4(4)(-3)}}{2(4)}$

   $x=\dfrac{-4\pm\sqrt{64}}{8}=\dfrac{-4\pm8}{8}=\dfrac{-12}{8},\dfrac{4}{8}=-\dfrac{3}{2},\dfrac{1}{2}$

5. a. $y-x\geq2$

$-x\geq2-y$

$x\leq y-2$

b. $y<x^2+2$

$y-2<x^2$

$\sqrt{y-2}<x$

$x>\sqrt{y-2}$

c. $\ln x>2-y$

$x>e^{2-y}$

d. $2-\dfrac{1}{x}\leq y^2$

$-\dfrac{1}{x}\leq y^2-2$

$\dfrac{1}{x}\geq2-y^2$

$x\geq\dfrac{1}{2-y^2}$

Chapter 9 Answers to Practice Exercises

1. a. $\$5,000,000=\$5\times10^6=5M\$$ (megabucks!)

b. $2,000,000,000$ bytes $=2\times10^9$ bytes $=2$ Gbytes or 2 Gb

c. 0.025 m $=2.5\times10^{-2}$ m $=2.5$cm

d. $4\times10^{-9}=4$ ns

e. 8×10^{-3} W $=8$ mW (milliwatts)

f. 10^{-6} m $=1$ μm, also called a "micron"

g. 50×10^3 tons $=50$ kton (kilotons)

h. 1×10^{12} dactyls $=1$ Tdactyl (a teradactyl!)

2. a. 5.3Mbar$=5.3\times10^6$ bar or $5,300,000$ bar

b. $0.2kJ=0.2\times10^{3}$ J=200J

c. $20mm=20\times10^{-3}$ m=0.020m

d. $1.1ng=1.1\times10^{-9}$ g=0.0000000011g

3. a. $(1.0yr)\left(365^{dy}\!\!/_{yr}\right)\left(24^{hr}\!\!/_{dy}\right)\left(60^{min}\!\!/_{hr}\right)\left(60^{s}\!\!/_{min}\right)=3.1536\times10^{7}$ s $=3.2\times10^{7}$ s

b. 5.29×10^{-11} m$=52.9\times10^{-12}$ m=52.9pm

c. $\dfrac{(500yd)\left(3^{ft}\!\!/_{yd}\right)}{5280^{ft}\!\!/_{mi}}=0.284mi$

d. $(0.0025mi)\left(5280^{ft}\!\!/_{mi}\right)\left(12^{in}\!\!/_{ft}\right)=158.4in$

$\dfrac{(158.4in)\left(2.54^{cm}\!\!/_{in}\right)}{\left(100^{cm}\!\!/_{m}\right)}=4.0m$

e. $\dfrac{(0.35lb)\left(453.6^{g}\!\!/_{lb}\right)}{\left(1000^{g}\!\!/_{kg}\right)}=0.16kg$

f. $1.2\mu s=1.2\times10^{-6}$ s$\left(\dfrac{10^{9}\ ns}{1\ s}\right)=1.2\times10^{3}$ ns

g. $101.3MHz=101.3\times10^{6}$ Hz$\left(\dfrac{1\ kHz}{10^{3}\ Hz}\right)=101.3\times10^{3}$ kHz

(kHz and MHz are measures of radio frequency)

h. $\dfrac{100g}{453.6^{g}\!\!/_{lb}}=0.220$ lb

4. a. 96485 C·mol^{-1}

b. 0.08206 atm·L·mol^{-1}·K^{-1}

c. 2.998×10^{8} m·s^{-1}

d. 1 kg·m^{-1}·s^{-2}

5. a. $1\,\dfrac{J}{s}$

b. $9.81\,\dfrac{m}{s^2}$

c. $8.314\,\dfrac{J}{mol \cdot K}$

d. $1\,\dfrac{kg \cdot m^2}{s^2}$

Chapter 10 Answers to Practice Exercises

1. a. This is not a straight line, there is a curved part that flattens out:

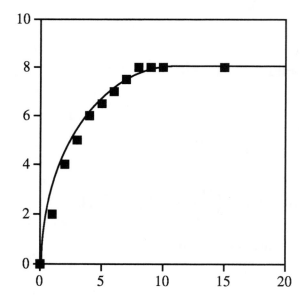

b. This is a straight line, which you can put through the points "by eye":

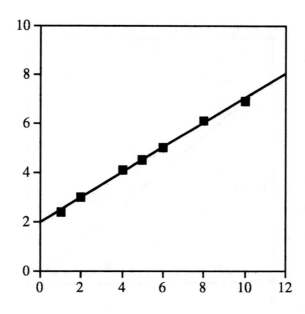

2. a. This line is drawn easily using the equation:

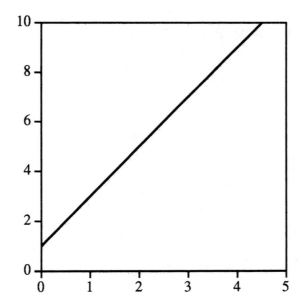

b. You need to rearrange the equation first to $y = -\frac{1}{2}x - 1$:

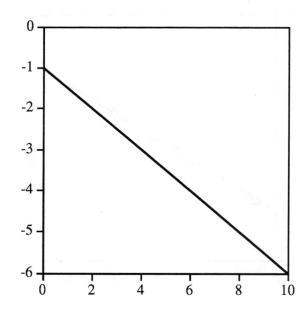

3. 1(b) The intercept is near 2, and the slope is near 0.5. The line follows the equation $y = \frac{1}{2}x + 2$.

2(b) The intercept is -1, and the slope is $-\frac{1}{2}$ (see answer to 2(b) above).

4. a. 3.5

b. 9

c. -7

d. 1.46

e. 0

f. -12

5. A table for Least Squares analysis looks something like this:

x	y	x^2	xy
1	2.4	1	2.4
2	3	4	6
4	4.1	16	16.4
5	4.5	25	22.5
6	5	36	30
8	6.1	64	48.8
10	6.9	100	69
Sum	36	32	246	195.1

$$m = \frac{7(195.1) - (36)(32)}{7(246) - (36)^2} = 0.502$$

$$b = \frac{(246)(32) - (36)(195.1)}{7(246) - (36)^2} = 1.992$$

Chapter 11 Answers to Practice Exercises

1. a. $\pi = 180°$

b. $\pi/4 = 45°$

c. $-3\pi/2 = -270° = 90°$

d. $2.43\left(\dfrac{180°}{\pi \text{ radians}}\right) = 139°$

2. a. $135° = 3\pi/4$

b. $-270° = -3\pi/2 = \pi/2$

c. $30°\left(\dfrac{\pi \text{ radians}}{180°}\right) = 0.5236$ radians

d. $257.3°\left(\dfrac{\pi \text{ radians}}{180°}\right) = 4.491$ radians

3. a. 0

b. 0

c. $\sin(45°) = \sqrt{\frac{1}{2}} = 0.70711$

d. 1.732

4. a. $\cos(23°) = \dfrac{2.5}{H}$ $H = 2.716$

b. $\cos\theta = \dfrac{A}{H} = \dfrac{3.0}{4.0} = 0.75$

$\theta = \cos^{-1}(0.75) = 41.4°$

c. $\tan\theta = \dfrac{O}{A} = \dfrac{4.4}{2.2} = 2.0$

$\tan^{-1}(2.0) = 63.4°$

$H = \dfrac{O}{\sin\theta} = \dfrac{4.4}{\sin\left(63.4°\right)} = 4.9$

d. $\sin(62°) = \dfrac{O}{5.0}$

Opposite $= 5.0\sin(62°) = 4.4$

$\cos(62°) = \dfrac{A}{5.0}$

Adjacent $= 5.0\cos(62°) = 2.3$

Angles $= 62°, 90°,$ and $90° - 62° = 28°$

5. For a–d, use the Pythagorean Theorem,

$H^2 = O^2 + A^2$

where H is the hypotenuse and O and A represent the other two legs of the triangle. In the solutions below, the third leg is represented by the variable L.

a. $(5)2 = (4)^2 + L^2$

$L^2 = 25 - 16 = 9$

$L = \sqrt{9} = 3$

b. $(4)^2 = (3)^2 + L^2$

$L^2 = 16 - 9 = 5$

$L = \sqrt{5} = 2.24$

c. $(H)^2 = (2)^2 + (1)^2 = 4 + 1 = 5$

  $H = \sqrt{5} = 2.24$

d. $(6)^2 = (3)^2 + L^2$

  $L^2 = 36 - 9 = 27$

  $L = \sqrt{27} = \sqrt{3 \cdot 9} = 3\sqrt{3} = 5.20$

6. a. $\sin^{-1}(0.8290) = 56.00°$

  b. $\cos^{-1}(-0.6018) = 127.0°$

  c. $\tan^{-1}(1.00) = 45°$

  $\sin(45°) = \sqrt{\frac{1}{2}} = 0.707$

  d. $\cos^2(45°) = (\cos(45°))^2 = (0.707)^2 = (\sqrt{\frac{1}{2}})^2 = 0.5$

7. a. $A_x = -4,\ A_y = 3,\ |A| = \sqrt{(-4)^2 + (3)^2} = \sqrt{25} = 5$

  $\theta = \tan^{-1}\left(\dfrac{3}{-4}\right) = -36.9°$

  b. $\overline{A} + \overline{B} = (-2,2) + (4,1) = (-2+4, 2+1) = (2,3)$

  c. $\overline{A} - \overline{B} = \overline{A} + (-\overline{B}) = (-2,2) + (-4,-1) = (-6,1)$

8. a. $P = 4(4.2\ cm) = 16.8\ cm$

  $A = (side)^2 = (4.2\ cm)^2 = 17.64\ cm^2 = 1.8 \times 10^1\ cm^2$

  b. $A = \pi R^2 = \pi\left(\dfrac{D}{2}\right)^2 = \pi\left(\dfrac{6.0\ in}{2}\right)^2 = \pi(3.0\ in)^2 = 28.27\ in^2 = 2.8 \times 10^1\ in^2$

  c. $V = L \times W \times D = 2 \times 4 \times 5 = 8 \times 5 = 40$

  d. $V = H \times \pi R^2 = (10.0)(\pi)(2.0)^2 = 1.3 \times 10^2$

  e. $V = \frac{4}{3}\pi R^3 = \frac{4}{3}(\pi)(1 \times 10^{-6}\ m)^3 = 4.2 \times 10^{-18}\ m^3$

  f. $A = \frac{1}{2}(B \times H) = \frac{1}{2}(10.0\ in)(2.00\ in) = 10.0\ in^2$

Chapter 12 Answers to Practice Exercises

1. $m = 22.0$ g

$V = 14.3$ cm^3 - 10.0 cm^3 = 4.3 cm^3

$$\text{Density} = \frac{\text{Mass}}{\text{Volume}} = \frac{22.0 \text{ g}}{4.3 \text{ cm}^3} = 5.1 \text{ }^{g}/_{cm^3}$$

This is not gold. Pyrite ("fool's gold") has a density near $5 \text{ }^{g}/_{cm^3}$.

2. $c = 2.998 \times 10^8 \text{ }^{m}/_{s} = \dfrac{\text{distance}}{\text{time}}$

time of pulse = 5 ns

Distance = (time)(c) = $(5 \times 10^{-9} \text{ s})(2.998 \times 10^8 \text{ m/s}) = 1.5$ m (about 5 ft)

For a shorter pulse, time = 2 ps = 2×10^{-12} s

Pulse length = $(2 \times 10^{-12} \text{ s})(2.998 \times 10^8 \text{ m/s}) = 6 \times 10^{-4}$ m = 0.06 cm = 0.02 in

3. R_{marble} = 5.0 mm

D_{marble} = 2(R) = 10 mm = 1.0 cm

D_{tire} = 10.0 ft

Circumference of tire = $\pi D = \pi(10.0 \text{ ft}) = 31.4$ ft

The number of marbles will be equal to $\dfrac{\text{Circumference}_{tire}}{\text{Diameter}_{marble}}$:

C_{tire} = (31.4 ft)(12 in/ft)(2.54 cm/in) = 957 cm

$\text{\# marbles} = \dfrac{C_{tire}}{D_{marble}} = \dfrac{957 \text{ cm}}{1.0 \text{ cm}} = 957$ marbles

4. v_{sound} = 1140 ft/s

distance = (5.0 mi)(5280 ft/mi) = 26400 ft

$$\text{Velocity} = \frac{\text{Distance}}{\text{Time}}$$

$$\text{Time} = \frac{\text{Distance}}{\text{Velocity}} = \frac{26400 \text{ ft}}{1140 \text{ }^{ft}/_{s}} = 23.2 \text{ s}$$

It is relatively easy to estimate seconds between a lightning strike and the thunder clap (count 1-Mississippi-2-Mississippi, etc.). An equation which gives a relation between each second counted and the distance would be handy:

Distance (in feet) = time (in seconds) $\times$ 1140 ft/s

Distance (in miles) = time (in seconds) $\times$ 0.22 mi/s

Since 0.22 is close to one fifth, a good estimate would be

$$\text{Distance (in miles)} = \frac{\text{Time (in seconds)}}{5}$$

5. h = 6.63 $\times$ 10^{-34} J·s

m = 150 g = 0.150 kg   (need kg for equation)

$$v = \frac{(90\ ^{mi}/_{hr})(5280\ ^{ft}/_{mi})(12\ ^{in}/_{ft})(2.54\ ^{cm}/_{in})}{(60\ ^{min}/_{hr})(60\ ^{s}/_{min})(100\ ^{cm}/_{m})} = 40\ ^{m}/_{s}$$

$$\lambda = \frac{h}{mv} = \frac{6.63 \times 10^{-34}\ J \cdot s}{(0.150\ kg)(40\ ^{m}/_{s})} = 1.1 \times 10^{-34}\ \frac{kg \cdot m^2 \cdot s^{-2} \cdot s}{kg \cdot m \cdot s^{-1}}$$

λ = 1.1 $\times$ 10^{-34} m, an extremely short wavelength

6. Area(wall) = 9 ft $\times$ 9 ft = 81 ft^2

Area(roll) = 1.5 ft $\times$ 20 ft = 30 ft^2

$$\#rolls = \frac{\text{Area(wall)}}{\text{Area(roll)}} = \frac{81\ ft^2}{30\ ft^2} = 2.7$$

You must buy three rolls.

7. Area(rink) = 200 ft $\times$ 50 ft = 10,000 ft^2

Density(ice) = 0.92 g/cm^3 = mass/volume

Need the volume of ice, determined from the area and thickness:

$$\text{Volume = Area} \times \text{Thickness}$$

Since the density is given in g/cm^3, we need to convert volume to units of cm^3. Calculate it in in^3 first, after converting the area to in^2:

Area = (10000 ft^2)(12 in/ft)^2 = 1.44 $\times$ 10^6 in^2

Note that the conversion factor was squared.

V = (1.44 $\times$ 10^6 in^2)(1.0 in) = (1.44 $\times$ 10^6 in^3)(2.54 cm/in)^3 = 2.36 $\times$ 10^7 cm^3

Mass = (0.92 g/cm^3)(2.36 $\times$ 10^7 cm^3) = 2.2 $\times$ 10^7 g = $\dfrac{2.2 \times 10^7\ g}{453.6\ ^{g}/_{lb}}$ = 4.8 $\times$ 10^4 lb

The ice weighs 48,000 pounds, or 24 tons.

8. Flour: $\dfrac{300 \text{ g}}{28 \text{ }^{g}/_{oz}} = 10.7$ oz

Butter: $\dfrac{180 \text{ g}}{28 \text{ }^{g}/_{oz}} = 6.4$ oz

Sugar: $\dfrac{70 \text{ g}}{28 \text{ }^{g}/_{oz}} = 2.5$ oz

Temperature:

$$T_{lower} = \left(\frac{9°\text{F}}{5°\text{C}}\right)(170°\text{C}) + 32°\text{F} = 338°\text{F}$$

$$T_{upper} = \left(\frac{9°\text{F}}{5°\text{C}}\right)(190°\text{C}) + 32°\text{F} = 374°\text{F}$$

Set the oven at around 350°F.

9. 14.7 psi = 14.7 lb air over 1 in² of earth

For 1 mi² base, $(5280 \text{ ft/mi})^2(12 \text{ in/ft})^2 = 4.01 \times 10^9 \text{ in}^2$

Set up proportion:

$$\frac{14.7 \text{ lb}}{1 \text{ in}^2} = \frac{x \text{ lb}}{4.01 \times 10^9 \text{ in}^2}$$

$x = (4.01 \times 10^9 \text{ in}^2)(14.7 \text{ lb/in}^2) = 5.9 \times 10^{10}$ lb of air

The maximum amount of argon is:

100% - 75.5% N_2 - 23.1% O_2 = 1.4% argon

The maximum weight of argon is thus

1.4%(5.9×10^{10} lb) = 0.014(5.9×10^{10} lb) = 8.3×10^8 lb = 410,000 tons argon

10. (10 furlongs)(10 chains/furlong)(100 links/chain)(7.92 in/link) = 79,200 in

$$\frac{79,200 \text{ in}}{12 \text{ }^{in}/_{ft}} = \frac{6600 \text{ ft}}{5280 \text{ }^{ft}/_{mi}} = 1\frac{1}{4} \text{ mile}$$

Chapter 13 Answers to Practice Exercises

1. Volume = 5 ft×5 ft×10 ft = 250 ft^3

$$\text{Density} = \frac{\text{Mass}}{\text{Volume}}$$

Mass = (Volume)(Density)

= (250 ft^3)(12 in/ft)^3(2.54 cm/in)^3(2.5 g/cm^3)

$\cong$ (250)(2000)(2.5)^3(2.5) g $\cong$ (250)(2000)(2.5)^2(2.5)^2 g

$\cong$ (500,000)(6)(6) g $\cong$ (1,000,000)(3)(6) g $\cong$ 18,000,000 g

$$\frac{18,000,000 \text{ g}}{454 \text{ g/lb}} \cong \frac{(900)(20,000)}{450} \cong 40,000 \text{ lb}$$

$$\frac{40,000 \text{ lb}}{2,000 \text{ lb/ton}} = 20 \text{ tons of bauxite}$$

2. Mass: $\dfrac{(158 \text{ grain})(454 \text{ g/lb})}{7000 \text{ grain/lb}} \cong \dfrac{(160)(450)}{(7000)}$ g

$\cong \dfrac{(16)(45)}{70} \cong \dfrac{(16)(9)}{14} \cong 9$ g bullet = 0.009 kg

Velocity: (2500 ft/s)(12 in/ft)(2.54 cm/in) $\cong$ (2500)(24 + 6) cm/s

$\cong$ (2500)(30) cm/s $\cong$ 75,000 cm/s $\cong$ 750 m/s bullet velocity

Energy: $E = \frac{1}{2}mv^2 \cong$ (0.5)(0.009 kg)(750 m/s)^2

$\cong$ (0.5)(0.009 kg)(75)(75)(10)(10) $\frac{\text{m}^2}{\text{s}^2}$ $\cong$ (0.5)(0.009)(75)(75)(100) $\frac{\text{kg·m}^2}{\text{s}^2}$

$\cong$ (0.5)(0.9)(75)(75) J $\cong$ (0.5)(75)(75) J   (round 0.9 to 1.0)

$\cong$ (0.5)(3)(25)(3)(25) J $\cong$ (0.5)(9)(625) J

$\cong$ (0.5)(6250) J $\cong$ 3120 J $\cong$ 3 kJ

(That would have been easier with a calculator)

3. pH = - log(0.004) = - log(4 × 10^{-3})

since log(5) = 0.7, estimate that log(4) $\cong$ 0.6, so

pH = -[log(4) + log(10^{-3})] $\cong$ -[0.6 - 3] $\cong$ -(-2.4) $\cong$ 2.4

4. $1 \text{ ft}^3 = (12 \text{ in})^3 = (144)(12) \text{ in}^3$

20 beans in 1 in² means the area $L^2 = 20$ beans, so $L = \sqrt{20} \cong 4.5$ beans

In 1 in³ there should be $(4.5)(20) \cong 90 \text{ beans/in}^3$

Total beans $\cong [(144)(12) \text{ in}^3][90 \text{ beans/in}^3]$

$\cong (144)(12)(90) \cong (144)(1000) \cong 144{,}000 \text{ beans}$ (that's a lot of beans)

5. a. One-way trip $\dfrac{1078 \text{ mi} \times 1.08 \text{ \$/gal}}{28 \text{ mi/gal}} \cong \dfrac{(1100)(1.1)}{28} \cong \dfrac{1100+110}{30} \cong \dfrac{1210}{30} \cong \dfrac{121}{3} \cong \$40$

A round trip would cost about \$80.

b. $\dfrac{1078 \text{ mi}}{65 \text{ mi/hr}} \cong \dfrac{1100 \text{ mi}}{65 \text{ mi/hr}} \cong \dfrac{220}{13} \cong \dfrac{225}{12.5} \cong 17 \text{ hr}$ one-way

17 hr + 4(20 min) = 18 hr 20 min, so a round trip = 36 hr 40 min

6. $\dfrac{2 \times 25 \text{ m} \times 100 \text{ cm/m}}{2.54 \text{ cm/in}} \cong \dfrac{2 \times 2500 \text{ in}}{2.5} \cong 2000 \text{ in per lap}$

(5280 ft/mi)(12 in/ft) = (5280)(12) in/mi

$\text{Laps} \cong \dfrac{(5280)(12) \text{ in/mi}}{2000 \text{ in/lap}} \cong (5.3)(6) \text{ lap/mi} \cong 30 + \tfrac{1}{3}(6) \cong 32 \text{ laps in a mile}$